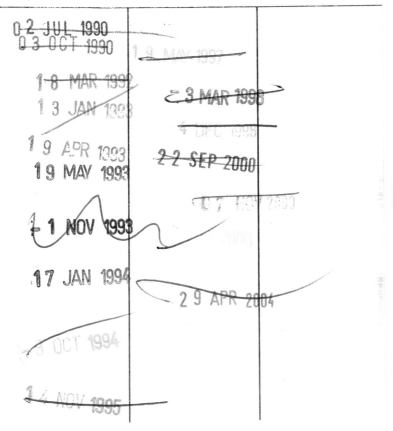

CARBON SUBSTRATES IN BIOTECHNOLOGY

Special Publications of the Society for General Microbiology

Publications Officer: Dr Duncan E.S.Stewart-Tull, Harvest House, 62 London Road, Reading RG1 5AS, UK

Publisher: Academic Press

1. Coryneform Bacteria

2. Adhesion of Micro-organisms to Surfaces

3. Microbial Polysaccharides and Polysaccharases

4. The Aerobic Endospore-forming Bacteria: Classification and Identification

5. Mixed Culture Fermentations

6. Bioactive Microbial Products: Search and Discovery

7. Sediment Microbiology

8. Sourcebook of Experiments for the Teaching of Microbiology

9. Microbial Diseases of Fish

10. Bioactive Microbial Products 2: Development and Production

11. Aspects of Microbial Metabolism and Ecology

12. Vectors in Virus Biology

13. The Virulence of Escherichia coli

14. Microbial Gas Metabolism

15. Computer-Assisted Bacterial Systematics

16. Bacteria in Their Natural Environments

17. Microbes in Extreme Environments

18. Bioactive Microbial Products 3: Downstream Processing

Publisher: IRL Press

19. Antigenic Variation in Infectious Diseases

20. Nitrification

21. Carbon Substrates in Biotechnology

This book is based on a symposium of the Fermentation and Physiology and Biochemistry Groups of the S.G.M. held in July 1986.

SPECIAL PUBLICATIONS OF THE SOCIETY FOR GENERAL MICROBIOLOGY
—————————— VOLUME 21 ——————————

CARBON SUBSTRATES IN BIOTECHNOLOGY

Edited by

J.D.Stowell

Pfizer Chemicals Europe and Africa,
10 Dover Road,
Sandwich CT13 0BN, UK

A.J.Beardsmore

Imperial Chemical Industries p.l.c.,
Biological Products Business,
PO Box 1, Billingham TS23 1LB, UK

C.W.Keevil

PHLS Centre for Applied Microbiology and Research,
Porton Down, Salisbury SP4 0JG, UK

J.R.Woodward

Director of Biotechnology,
University of Leeds,
Leeds LS2 9JT, UK

1987
Published for the
Society for General Microbiology
by

◇ IRL PRESS

Oxford·Washington DC

IRL Press Limited
PO Box 1
Eynsham
Oxford OX8 1JJ
England

British Library Cataloguing in Publication Data

Carbon substrates in biotechnology. — (Special publications of the
Society for General Microbiology; ISSN 0197 − 1751; no. 21)
1. Biotechnology   2. Carbon − Industrial applications
I. Stowell,J.D.   II. Series
660'.6   TP248.2

ISBN 1-85221-021-4 (hardbound)
ISBN 1-85221-020-6 (softbound)

Cover illustration. The cover design is from Figure 5 of Chapter 8 and shows a continuous oil sterilization system.

Printed by Information Printing Ltd, Oxford, England

Preface

The chapters in this volume are based on a symposium of the Fermentation and Physiology and Biochemistry Groups of the Society for General Microbiology, held at Durham University and ICI Billingham. The meeting was the fourth in a series organized by the group having joint academic and industrial hosts, the first three being held under the general heading of 'Bioactive Microbial Products'. The format was designed to bring together a multidisciplinary group of scientists engaged in fermentation biotechnology and to encourage an informal interchange of ideas and experiences. The formal presentations were seen as providing a framework for this.

We are very much indebted to the speakers, who not only gave stimulating presentations, but also participated in the lively discussions which followed and provided the manuscripts for this publication. We gratefully acknowledge the efforts of our colleagues at the University of Durham and ICI Billingham, who co-hosted and sponsored the meeting. Particular thanks are due to Derek Ellwood who welcomed the group to Durham University and introduced the symposium, and to Peter Rodgers who gave us an insight into the operation of the pilot plant facilities at ICI Billingham. Jill Seegers, Meetings Assistant of the Society for General Microbiology, is to be congratulated for her significant contribution to the smooth running of the meeting.

The subject of carbon feedstocks for biotechnology has achieved significant media coverage during recent years. Attention has focussed almost exclusively on the penalties incurred by fermentation manufacturers using carbohydrate feedstocks within the European Economic Community, a consequence of legislation originally designed to stimulate the agricultural economy. Indeed, one could be forgiven for concluding that carbohydrates are the only carbon feedstocks available for biotechnological processes. This is certainly not the case. New EEC legislation has now been announced which will address the problems facing local manufacturers and it therefore seemed timely to review the current position and future prospects for fermentation-based manufacturers within the Community.

Whilst, as noted above, carbohydrates are undoubtedly the best known fermentation feedstocks, many industrial processes are currently using alternatives and much research is being devoted to identifying potential new materials. Many biotechnological processes are competing with synthetic chemical or agricultural routes for a given product and the cost effectiveness of new production methods is vital. Overall process economics combine with quality considerations to dictate the choice of substrate for a particular process. The relative contribution of raw materials cost to overall product cost tends to increase as the production volume increases. It is the higher volume products which have become the target of many biotechnology companies and hence the importance of a correct approach to carbon substrates can hardly be overstated.

A long term view needs to be taken. For example, it is unsatisfactory for the competitiveness of an antibiotic producer to be subject to vagaries in the purchase price of a single agricultural commodity. Technical flexibility must be built in to processes and to this end an ongoing programme of research and development work is essential. A key objective of this symposium was to illustrate the range of carbon substrates available for biotechnology and to describe some of the special points associated with their use.

Many industrialists are undoubtedly unaware of much of the biochemistry surrounding the utilization of their chosen carbon substrates. Equally, there is little doubt that such a knowledge is a powerful tool in process development. A further objective of the meeting was, therefore, to bring together industrialists and academics, to outline selected aspects of the biochemistry of carbon utilization and hence to promote a cross-fertilization of ideas.

It would clearly be impossible to review comprehensively in a two-day meeting a subject as broad as carbon substrates in biotechnology. The aim was, rather, to present a broad introduction. If the quality and enthusiasm of the discussion can be taken as a measure then the meeting itself may be judged a success. It is hoped that this volume captures the flavour of the symposium and if the reader is stimulated to evaluate new approaches to the development of his or her process then the efforts will have been worthwhile.

J.D.Stowell

Contributors

C.Anthony
University of Southampton, School of Biochemical and Physiological Sciences, Bassett Crescent East, Southampton SO9 3TU, UK

J.Colby
Microbial Technology Group, Department of Microbiology, The Medical School, Framlington Place, Newcastle upon Tyne NE2 4HH, UK

S.H.Collins
Delta Biotechnology Ltd, 2nd Floor, Castle Court, Castle Boulevard, Nottingham NG7 1FD, UK

J.Coombs
Bioservices, London and Biotechnology Affiliates, Reading, UK

C.A.Cormack
CIBE, 29 Rue du Général Foy, F-75008 Paris, France

J.W.Drozd
Fermentation and Microbiology Division, Shell Research Ltd, Sittingbourne Research Centre, Sittingbourne, Kent ME9 9AG, UK

P.S.Gray
Commission of the European Communities, Directorate - General Science, Research and Development, Rue de la Loi 200, 1049 Bruxelles, Belgium

P.J.F.Henderson
Department of Biochemistry, University of Cambridge, Tennis Court Road, Cambridge CB2 1QW, UK

G.W.Logan
Microbial Technology Group, Department of Microbiology, The Medical School, Framlington Place, Newcastle upon Tyne NE2 4HH, UK

C.M.Lyons
Microbial Technology Group, Department of Microbiology, The Medical School, Framlington Place, Newcastle upon Tyne NE2 4HH, UK

M.C.J.Maiden
Department of Biochemistry, University of Cambridge, Tennis Court Road, Cambridge CB2 1QW, UK

O.M.Neijssel
Laboratory for Microbiology, University of Amsterdam, Amsterdam, The Netherlands

J.R.Quayle
The University, Claverton Down, Bath BA2 7AY, UK

J.D.Stowell
Pfizer Chemicals Europe and Africa, 10 Dover Road, Sandwich, Kent CT13 0BN, UK

H.Streekstra
Laboratory for Microbiology, University of Amsterdam, Amsterdam, The Netherlands

M.J.Teixeira de Mattos
Laboratory for Microbiology, University of Amsterdam, Amsterdam, The Netherlands

D.W.Tempest
Department of Microbiology, University of Sheffield, Sheffield, UK

E.Williams
Microbial Technology Group, Department of Microbiology, The Medical School, Framlington Place, Newcastle upon Tyne NE2 4HH, UK

J.Woodward
Chemical Technology Division, Oak Ridge National Laboratory, Oak Ridge, TN 37831, USA

Abbreviations

BHA	butylated hydroxyanisole
BHT	butylated hydroxytoluene
BIOPOL	biopolymer
CAP	Common Agricultural Policy
CAS	Concanavalin A – Sepharose
CCT	Common Customs Tariff
CEFIC	European Council of Chemical Manufacturers' Federations
CIBE	International Confederation of European Beet Growers
c.i.f.	cost-insurance-freight
DHA	dihydroxyacetone˙
DHAP	dihydroxyacetone phosphate
ECU	European Currency Unit
EFTA	European Free Trade Association
EtOH	ethanol
FAO	Food and Agriculture Organization
FBP	fructose bisphosphate
FPU	filter paper units
GSH	glutathione
HFCS	high fructose corn syrup
HLB	hydrophilic – lipophilic balance
ISA	International Sugar Agreement
KDPG	2-keto 3-deoxy 6-phosphogluconate
MDH	methanol dehydrogenase
MMO	methane mono-oxygenase
MMS	methyl methane sulphonate
MNNG	N'-methyl-N'-nitro-N-nitrosoguanidine
MSG	monosodium glutamate
Mt	million tonnes
MTBE	methyl-tertio-butyl ether
ORNL	Oak Ridge National Laboratory
PDH	pyruvate dehydrogenase
PEG	polyethylene glycol
PEP	phosphoenol pyruvate
PGA	phosphoglyceric acid
PHB	poly-β-hydroxybutyrate
p.m.f.	proton motive force
PQQ	pyrrolo-quinoline quinone
RuBP	ribulose bisphosphate
RuMP	ribulose monophosphate
SBP	sedoheptulose bisphosphate

SBPase	sedoheptulose bisphosphatase
SCOPA	Seed Crushers and Oil Processors Association
SCP	single cell protein
SV	saponification value
TA	transaldolase
TBA	tertio-butyl acid

Contents

CHAPTER 1

Impact of EEC regulations on the economics of fermentation substrates

P.S.GRAY

Commission of the European Communities, Directorate-General Science, Research and Development, Rue de la Loi 200, 1049 Bruxelles, Belgium

Introduction

The economics of carboyhydrate fermentation substrates are inextricably linked with agricultural and fiscal policy, the consumer of potable alcohol in its many forms being a major contributor to State budgets. The modern fermentation industry inherits a system which was conceived in a period when the great diversity of its products currently available did not exist and is thus ill-tuned to the needs of innovative biotechnology. It is currently fashionable to deride the European Community Common Agricultural Policy (CAP) which is now beset by burgeoning surpluses, but it is well to remember that at its inception a prophet of surplus would have been mocked as being over-gifted with imagination. England, and subsequently the United Kingdom, had similar systems of price and import controls in various forms from the Norman Conquest until the repeal of the Corn Laws in the nineteenth century. The aims of the CAP are expressed in Article 39 of the Treaty, namely:
The objectives of the Common Agricultural Policy shall be:

(a)   to increase agricultural productivity by promoting technical progress and by ensuring the rational development of agricultural production and the optimum utilization of the factors of production, in particular labour;

(b)   thus to ensure a fair standard of living for the agricultural community, in particular by increasing the individual earnings of persons engaged in agriculture;

(c)   to stabilize markets;

(d)   to assure the availability of supplies;

(e)   to ensure that supplies reach consumers at reasonable prices.

When decisions were taken as to how to achieve these objectives, there was a choice between guaranteed prices and deficiency payments. The Community of Six chose the former mechanism, the absence of the UK which operated a deficiency payments scheme being a crucial factor. However, the Community bio-industries faced a major handicap as a result of the difference in raw materials prices within the community and on the world market. Chemical and pharmaceutical companies situated in European Free Trade Association (EFTA) countries were able to purchase sugar — often of Community origin — at world prices in the range of 150−200 European Currency Units (ECU) per tonne, whereas Community manufacturers had to pay 550−580 ECU per tonne. Furthermore,

1

the absence of agricultural compensation measured against imports of the finished products rendered Community manufacturers uncompetitive in their domestic markets. For this reason, there was a continuing and recently accelerating shift of investment outside the Community.

Similarly, the difference in starch prices between the Community and its major competitor in the area of biotechnology, the United States, was of the order of 40%.

This situation not only deterred investment in the Community but also discouraged further research in these vital areas for future development. The Community therefore faced the danger of losing an important opportunity for replacing petrochemicals by indigenous natural renewable resources and for maintaining its competitive position in a new technology.

The manufacture of chemicals (including pharmaceuticals) from agricultural raw materials offered a major opportunity for the expansion of biotechnology in Europe at the same time as offering a growing outlet for agricultural produce. This opportunity is particularly great for the higher-priced speciality chemicals to which the European chemical industry is increasingly turning in the face of growing competition in bulk chemicals from new plants situated in the Middle East.

Sugar and starch are already the raw materials of choice for the manufacture of antibiotics, vitamins, enzymes, organic and amino acids. They can also be used to make certain lipids, solvents and plastics. Indeed, it would be technically feasible to base the whole chemical industry on agricultural feedstocks.

The chemical and pharmaceutical industries estimated that consumption of starch can increase to 1.6 million tonnes in 2000 (that is +400% over 1985) and of sugar to 550 000 tonnes (that is +580% over 1985), provided that raw materials were made available at world prices.

Faced with the problem, the Community set about reforming the agricultural regimes in order to put its bio-industries on an equal footing with those in third countries. This chapter analyses the situation prior to the reform and explains its consequences.

However, in order that the recently-adopted proposals for starch and sugar for industrial use can be fully understood, it is necessary to explain and to trace the history of the cereals and sugar regimes.

The cereals regime

The market for each cereal is governed by a reference price fixed by the agricultural ministers before the beginning of each campaign year. For practical purposes, there are two derived prices which are significant; the threshold price, which is the price at which the cheapest grain is allowed into the Community, and the intervention price. The threshold price for common wheat is fixed so that grain entering the Community at Rotterdam, once unloaded and transported by river to Duisburg (FRG), the point of greatest deficit, does not arrive below the reference price, and the intervention price is fixed by assuming that Orleans/Ormes in France is the point of greatest surplus, then by deducting the Orleans – Duisburg transport cost from the reference price. The difference between the intervention price at which the Community buys in surplus grain and the threshold price for common wheat is known as the margin of Community preference. In a deficit situation, prices rise to threshold as imports are price-determining. In a surplus situation, prices fall to or even below intervention since there is usually

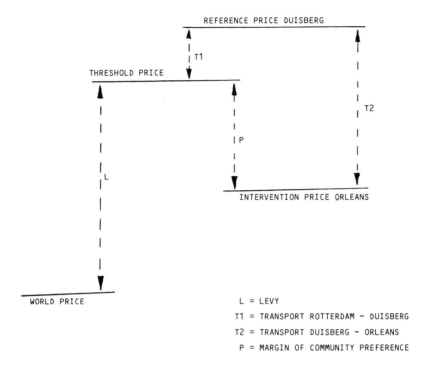

REFERENCE PRICE DUISBERG

THRESHOLD PRICE

INTERVENTION PRICE ORLEANS

WORLD PRICE

L = LEVY
T1 = TRANSPORT ROTTERDAM – DUISBERG
T2 = TRANSPORT DUISBERG – ORLEANS
P = MARGIN OF COMMUNITY PREFERENCE

Fig. 1. A schematic representation of the market regime for common wheat.

a payment delay. The system is shown schematically in *Figure 1*.

Farmers are encouraged to stock through the season by adding each month a storage charge of 2.57 ECU for common wheat to both intervention and threshold prices. The impact of this scheme on the maize market, which is subject to a similar regime, is shown in *Figure 2*, the saw edge annual fluctuation being due to the storage charge which encourages farmers to hold stocks on the farm rather than sell early in the campaign year.

Imported wheat is brought up to threshold price by imposing a levy which is calculated and published daily on cost-insurance-freight (c.i.f.) offers of wheat at Rotterdam. The catch is that the levy, whilst calculated on the lowest price offer, is applied to all imports whatever the purchase price on the world market, so that the importers of high quality more expensive North American wheat will pay a landed price net of levy well over the threshold price. Exports from the Community get a refund which, for grain, is struck weekly by means of a tender procedure.

This system would have disadvantaged the Community food industry where its competitors had access to cheaper raw materials, and therefore a parallel system of variable components of duty (levies) and export refunds was introduced for most food, and for some non-food, products made from cereals and other raw materials subject to a price regime. These variable components and refunds were calculated on the raw material content of the processed foodstuff. Thus, since 1.67 tonnes of durum wheat are required to make 1 tonne of pasta, the variable component on pasta is 1.67 times the levy on one tonne of durum wheat (Anon, 1980).

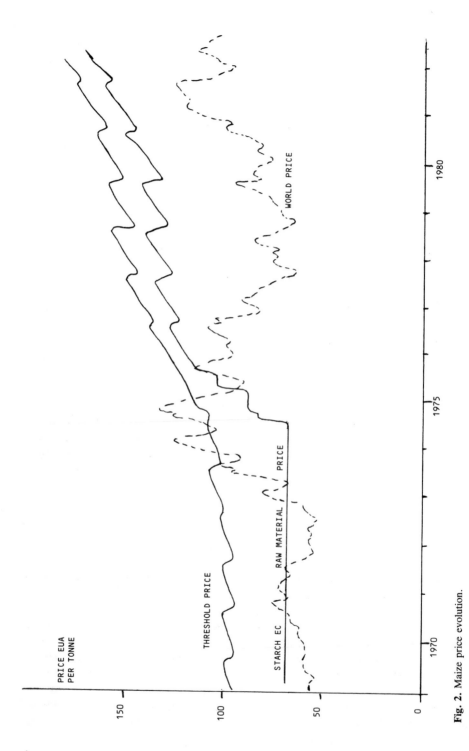

Fig. 2. Maize price evolution.

Starch

It was recognized that starch manufacturers suffered some disadvantages from the market management system and therefore a system of production refunds was introduced, with the original intention of bridging the gap between world and Community prices for the starch manufacturers. Whilst at the outset this was more or less true, as time progressed and Community prices drew away from world prices, the production refund became inadequate, so that a competitive gap opened up between producers inside the Community who were linked to the internal price regime and external manufacturers who could buy raw materials at world prices. *Figure 2* shows the change in competitive advantage from 1970 to 1982 and *Figure 3* the actual situation in 1967 and 1981 for maize.

The reason for this is not far to seek since agricultural Councils had to decide each year the new level for the production refund and, as the cost of the CAP grew, there was a tendency to concentrate spending on price support rather than on spending which could be seen as a compensation for industry. The effect of this failure to adjust adequately the starch production refund can be seen in *Figure 3*.

Sugar

In 1978 a regime for sugar for the chemical industry was introduced and here refunds were paid to the manufacturers of a list of products. The refund paid was twice the starch refund per tonne of sugar, and so the compensation was linked to the starch refund and suffered all its disadvantages in not following the differences between world and internal market prices. It also suffered a further disadvantage in that the list of products was limited and precisely defined in the Annex to the Regulation (Anon, 1978).

Additions to this list, which were necessary when new products were developed, were made by Council decision on a proposal from the Commission — a cumbersome and time-consuming procedure. A manufacturer wishing to develop a new product was not sure from the outset whether he could benefit from a refund or not, and once his application was filed his intentions were revealed to his competitors. In practice, this meant that manufacturers preferred to site plants for new products outside the Community rather than run the risks associated with divulging their production strategy, and that with no certainty of benefiting from the refund.

Industrial consequences of the agricultural policy

The choice of raw material for a biotechnological process will usually be decided largely on economic grounds. Although there are technical differences in different raw materials, the implications of these are usually relatively minor. Indeed, in the US, many processes are capable of running on a range of raw materials, choice being made on the ruling substrate price and availability. In Europe, however, such multi-substrate systems are rare. Whilst different efficiencies are expected from different raw materials, this is often a reflection of the concentration of research and development effort on the cheaper raw materials. Terramycin can be produced (and is in the US) from both vegetable oil and sucrose; citric acid can be produced from vegetable oil, molasses, hydrocarbons and dextrose.

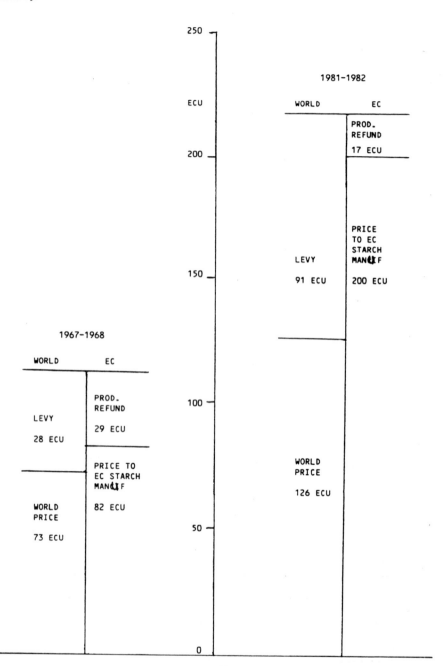

Fig. 3. The change in competitive advantage for starch (maize).

The effect of raw material costs on manufacturing costs is illustrated in *Table 1*.

The effect of the market regimes and import levy and refund system can be best illustrated by taking the case of a simple product, citric acid. There was no variable component of duty levied on citric acid imported from outside the Community. The case

6

R = EC RAW MATERIAL COST
r = NON-EC RAW MATERIAL COST
E = EXPORT REFUND
D = DUTY ON CITRIC ACID (13.6%)
M = MANUFACTURING COST
P = SUGAR PRODUCTION REFUND

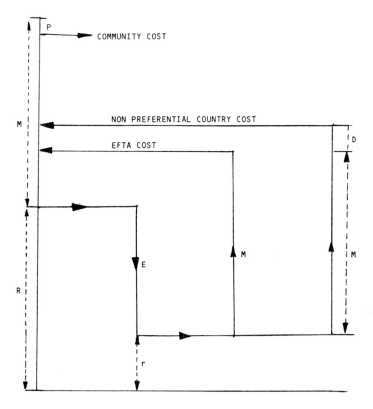

Fig. 4. Economics of citric acid production.

Table 1. Biotechnological processes — raw material costs.

Product	Value ($/tonne)	Raw material costs (%)
Bacterial protein	383	74
Fungal protein	919	55
Molasses yeast	223	46
Insecticide	875	37
Bacillus thuringiensis	4185	42
Monosodium glutamate (MSG)	664	35
Antibiotic	15200	32

is illustrated diagrammatically in *Figure 4*. A manufacturer living in a neighbouring state such as one of the EFTA countries could import Community sugar under bond. As the sugar left the Community, it benefited from a refund R which, in the case illustrated, is assumed to be 70% of the Community price. The sugar was then converted into citric acid and it is assumed, for the sake of illustration, that the raw materials represented 50% of the manufacturing cost. On re-importing to the Community, no variable component of duty was levied, so that the external EFTA-based manufacturer could market his product at a competitive advantage of over 30% less than the EEC manufacturer due to raw materials alone. Even if the manufacturer was in a non-preferential third country and had to pay fixed duty, he could still enjoy a large competitive advantage.

The refund for sugar for the chemical industry redressed the balance a little, but an industrialist making decisions about his investment strategy was still faced with a number of uncertainties.

(i)   The evolution of the starch/sugar refund which was adjusted at each agricultural price fixing.

(ii)   Whether a new product would be accepted as qualifying for a refund by a rather lengthy decisional procedure.

(iii)   Commodity price fluctuations on the world market.

(iv)   The dollar/ECU monetary relationship.

New policy options

The aim of the change in existing agricultural regimes in relation to the bio-industries was to eliminate the competitive disadvantages faced by Community industry on its home market by making available agricultural feedstocks at prices equivalent to those prevailing on the markets of main third country competitors. In the absence of agricultural levy protection against imports, the Community bio-industry had to be able to obtain raw materials at world market conditions.

While it was recognized that industry would have to accept a degree of uncertainty about fluctuations in world market prices, it was essential that the system be as transparent and stable as possible so that companies are able to plan ahead.

A number of options were open to the Community. The simplest of these would seem to have been to pursue the existing regime by completing the list of products for which variable components of duty were imposed and improving the procedure for listing new products for the production refund. This approach would, however, have involved imposing new variable components which would have to be paid for by concessions in other fields to our trading partners. In the current economic climate it would have been difficult to find concessions which could be offered in compensation. Whatever improvements would have been made for adding products to a positive list, there would still have been a degree of uncertainty as to whether and when a new product could have qualified for a refund. Manufacturers would have been discouraged not only from undertaking research because of the uncertainties surrounding raw material price calculations but also from investing in manufacturing capacity in the Community, knowing that both the original decision on the granting of a production refund and the amount of refund would be subject to political uncertainty.

It was clear that if research and investment were to be encouraged and new outlets opened for agricultural raw materials, then Community industry must be put on the same competitive basis as those of our trading partners, not only on the export but also on the internal market. Since the new biotechnology industry could not be accommodated in the existing mechanisms these new mechanisms would have to be created.

In its Communication on Agricultural Policy in 1983 (Anon, 1983a) the Commission stated its intention to propose to adjust cereals prices towards world prices. This would mean that over a period, as far as starch was concerned, the competitive disadvantage should disappear as EEC and world prices moved closer together.

In its Communication on Biotechnology in 1983 (Anon, 1983b) the Commission suggested that raw materials for fermentation into industrial products should be made available to industry at world prices. This statement was reinforced in the EEC Green Paper on Agricultural Policy in 1985 (Anon, 1985). If these two policy statements were combined, then it would be necessary for the Community to finance the raw materials regime with respect to starch for a limited period only until Community and world prices were aligned. The cost of covering this interim period could therefore be looked on as an incentive payment to launch the new industry. As far as sugar was concerned, the thinking went through a number of changes. The first proposal was that sugar for industrial products in excess of the current usage should be C sugar, namely sugar which was normally only to be sold outside the Community, but which would now be available for sale at a price to be negotiated with the user with world price as a barometer. This proposal suffered from a number of disadvantages.

Firstly, C sugar was a child of fortune, its production depending on chance as much as planning since there was no guaranteed outlet. Secondly, the availability of C sugar in the Member States did not necessarily match up to industrial demand. Thirdly, the existence of three sugar prices on the internal market — a sugar for food use, a quota for each chemical based on current use of the old sugar for the chemical industry price and finally a price based on C sugar for extra usage. This variety of prices would have made accounting difficult and fraud attractive.

Following a series of proposals from the Commission and negotiation in the Council, new regimes have been agreed for starch and sugar, the basic elements of which are:

(i)   world price for industrial use;
(ii)   a fairly rapid run-in period;
(iii)   an extensive list by chapter tariff headings covering all principal groups of industrial products;
(iv)   a breaking of the link between sugar and starch;
(v)   the payment of the refund (price compensation) to the user rather than to the raw material producer.

The new regimes were adopted on 25 March 1986 and came into application on 1 July of that year.

Scope of the new regimes

As far as fermentation substrates are concerned, the products which are eligible for beet starch and sugar refunds are identical and are listed in *Table 2*.

The starch regime has additional beneficiary products in the textile and paper sector where native starch is used for purposes other than as a fermentation substrate. As can be seen, the list in *Table 2* has very broad headings, for example Chapter 30, phar-

Table 2. Products for which production refunds are given.

Common Customs Tariff (CCT) heading No.	Description
ex 13.03 C III	Carrageenan
ex 15.11 B	Glycerol, other than crude
Chapter 29 [excluding subheadings 29.04 C II, 9.04 C III and ex 29.43 B (levulose)]	Organic chemicals
Chapter 30	Pharmaceutical products
34.02	Organic surface-active agents; surface-active preparations and washing preparations, whether or not containing soap
Chapter 35 (excluding headings 35.01 and 35.05)	Albuminoidal substance; glues; enzymes
Chapter 38 (excluding subheadings 38.12 A and 38.19 T)	Miscellaneous chemical products
Chapter 39	Artificial resins and plastic materials, cellulose esters and ethers, articles thereof

maceutical products. Thus, the manufacturer of a new pharmaceutical product knows from the outset that he will receive a refund for his product and can thus make cost calculations based on world prices for raw materials.

The starch regime

The refund applicable in the 1985–1986 marketing year was granted to all starch, whatever its destination. Since the object of the new regime is to give full compensation where products are not protected by the import/export regime for processed products, and not to compensate where there is trade protection, the first element of the regime dealt with in Article 4 of the Regulation (Anon, 1986a) is a 3-year phasing-out of the current refund system by a 25% reduction each year, the first reduction being made on 1 July 1986.

To encourage rapid development of the biotechnological industries, the new regime for non-protected products will be introduced in two steps. In the first step, the unprotected products will be given 75% of the old refund plus 50% of the difference between this figure and the 'new' refund. In the second year the unprotected products will be given a refund equal to the 'new' refund. By 'new' refund we mean a refund which should give starch manufacturers their raw materials at world prices. There are, however, three principal raw materials used for starch manufacture — potatoes, maize and wheat. The economics of these industries are very complex and depend on a number of factors, including pollution problems and by-product returns.

Faced with this problem, the Council decided to give a premium for starch potatoes and give one single refund based on the difference between the common intervention price for cereals and the world price as determined by c.i.f. Rotterdam prices. The details of this are still being worked out in the implementing regulations which, at the time this chapter was being written, were not adopted. Many factors will affect the evolution of starch production costs in the next few years, and among these are the

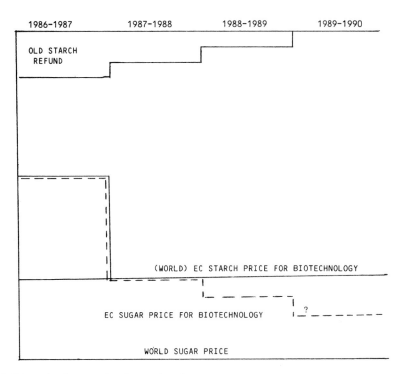

Fig. 5. Schematic of new regime for starch and sugar.

following:

(i)   The price of wheat in a market which has an inbuilt structural surplus.

(ii)   The price of wheat which is unsuitable for intervention where intervention quality is being continually improved.

(iii)   The return on by-products such as vital gluten and corn gluten feed.

(iv)   The evolution of world prices which are quoted in dollars and the dollar – ECU exchange rate.

The principles of the regime, however, are shown diagrammatically in *Figure 5*.

The sugar regime

The new sugar regime allows the industrial user access to sugar at world price plus a standard amount of 70 ECU/tonne which is estimated to be the cost of transporting white sugar from other production sites into the Community (Anon, 1986b). The world market price is derived by deducting the average export refunds from Community sugar price. There is, however, a transitional period and in the 1986/87 and 1987/88 marketing years the sugar price is deemed to be equal to the glucose price derived from starch. That is to say that during this period the refund will maintain both starch and sugar on an *ex aequo* basis. However, in the following two years the sugar price will be adjusted downwards by 25% of the difference between the glucose and world sugar price in 1988/89 and 50% of the difference in 1989/90 with a requirement in Article 4(3)(b) that, if the aid for sugar disrupts the starch sector, then the adjustment shall not be

11

raised to 30% and the arrangement shall be reviewed.

Definitions of the Community price for sugar and the way in which the glucose price is to be calculated are also included in Article 4. The regime is also shown diagrammatically in *Figure 5*.

Concluding remarks

The need for brevity in this chapter may have created the impression that the new regimes have created more difficulties than they solve. The fact of the matter is that market regimes exist for the major outlets for sugar and cereals, namely the food sector. The sugar regime is self-regulatory in that production is planned and an annual balance drawn between internal demand, export possibilities, prices and the cost of refunding exports. The balance is expressed in the form of fixed quotas. The cereals regimes do not have fixed quotas but a price support mechanism and conditions for import and intervention.

Bringing fermentation substrates to world price level involves fencing off a part of the CAP from the operation of the market regimes, so that process can be no less complex than the regimes themselves. It may be that the detailed operation has to be tuned, but the main message is that the European Community has recognized the importance of the fermentation industries as increasing users of agricultural raw materials and creators of wealth and employment, it is now up to industry to respond to this political act of confidence.

References

Anon (1978) *Council Regulation (EEC) No. 1400/78*, OJ No. L 170 of 27 June 1978, p. 9.

Anon (1980) *Council Regulations (EEC) Nos 3033/80, 3034/80, 3035/80*, OJ No. L 323 of 29 November 1980, pp 1–50.

Anon (1983a) *COM(83)500* final of 28 July 1983.

Anon (1983b) *COM(83)328* final of 8 June 1983.

Anon (1985) *COM(85)333* final of 13 July 1985.

Anon (1986a) *Council Regulation (EEC) No. 1009/86*, OJ No. L 94 of 4 April 1986, p. 6. See also Regulations (EEC) Nos 1006/86–1008/86 in the same Official Journal, pp 1–5.

Anon (1986b) *Council Regulation (EEC) No. 1010/86*, OJ No. L 94 of 9 April 1986, p. 9.

CHAPTER 2

Arable crops for the fermentation industry — a grower's view point

Deputy Director of the International Confederation of European Beetgrowers, 29, rue du Général Foy, F-75008 Paris, France

Introduction

The Common Agricultural Policy has more than proved its worth in meeting the food requirements of the population of the European Economic Community. Traditional food outlets within the Community are now virtually saturated and those outside only offer limited growth. These are the factors which have driven arable farmers to seek new non-food outlets for their products and the fermentation industry is one of these. This chapter charts this development and explores the implication of a new non-food agriculture for arable farming.

The establishment of the Common Agricultural Policy

The first mention of a common policy for agriculture comes in Article 39 of the Treaty of Rome (1957) establishing the European Economic Community. It has to be remembered that these objectives were established in the context of the post-war period of food scarcity and dependence upon imports of certain strategic commodities. These objectives were as follows.

(i)   To increase agricultural productivity by promoting technical progress and by ensuring the rational development of agricultural production and the optimum utilization of the factors of production, in particular labour.
(ii)   Thus to ensure a fair standard of living for the agricultural community, in particular by increasing the individual earnings of persons engaged in agriculture.
(iii)   To stabilize markets.
(iv)   To assure availability of supplies.
(v)   To ensure that supplies reach consumers at reasonable prices.

The CAP has been successful in largely meeting these objectives, particularly with respect to the availability of food supplies, since self sufficiency for most agricultural products has now been achieved (see *Table 1*).

Table 1. Degree of self-supply for the principal agricultural products (%).

	1964	1973	1982	1990
Cereals	75	90	105	127
Meat	94	92	100	100
Sugar	103	92	154	122
Milk	100	99	103	102
Wine	98	90	94	123

From Community Agricultural Situation, 1984 Report (Anon, 1984a).

Table 2. The estimated growth of the EEC population (1000).

	1983	1990	2000
Belgium	9856	9887	9972
Denmark	5114	5061	4940
Germany	61 423	60 640	59 143
Greece	9847	9880	10 435
France	54 729	56 139	58 573
Ireland	3508	3799	4247
Italy	56 836	57 331	57 966
Portugal	9946	10 577	11 053
Spain	38 173	39 635	41 117
Netherlands	14 367	14 973	15 643
United Kingdom	56 377	56 785	57 902
EEC 12	320 541	325 077	331 364

From Eurostat − 23rd edition, 1985 (Anon, 1985a).

Indeed, for certain agricultural products, the level of supply has well exceeded the required level of self-sufficiency, as is the case for sugar, cereals and milk. There is nothing inherently wrong with exceeding levels of self-sufficiency providing there is a realistic possibility to export that which is surplus to domestic requirement.

The imbalance between food supply and demand in the EEC

The imbalance between the level of demand and supply of agricultural products is a result of several factors. Essentially, there is a contrast between the dynamic growth of food supply from the agricultural sector and the slow growth, or even decline, in demand for agricultural products from the consumer. European agriculture is now confronted by the logistical barrier of the saturation of its domestic food markets. Weak population dynamics are one explanation for this slow growth in demand (see *Table 2*).

The population of the EEC had an annual rate of growth of 0.8% in the 1960s, but this slowed to 0.4% in the 1970s and has now fallen to 0.2% per annum at which level it is likely to remain well into the 1990s. The inclusion of Spain and Portugal into the Community has raised the total population from 272 to 320 million, but the rate of growth of the population will probably not be increased. Furthermore, these two countries bring agricultures with a great potential to expand.

Other factors have strongly influenced the growth of demand, notably the general economic situation which has an impact on the purchasing power of consumers. There

Table 3 Changes in the consumption of certain food products in the EEC (kg/capita).

	1973/74	*1982/83*
Cereals	85	83
Sugar	37	34
Vegetables	100	107
Fruit	60	59
Wine	50	45
Milk	97	102
Butter	5	6
Meat	75	88

From Community Agricultural Situation, Report 1985 (Anon, 1985b).

Table 4. Yield development of certain arable crops in the EEC (tonnes/hectare).

	1961/64	*1981/84*
Wheat	3.3	5.6
Barley	3.2	4.4
Maize	2.8	6.5
Potatoes	21.3	32.0
Sugar beet	37.8	52.0

From Eurostat 23rd Edition 1985 (Anon, 1985a).

are some signs of recovery from the general economic recession: the rate of economic growth for the Community increased from 2.2% per annum in 1984 to 2.3% in 1985 and the overall rate of inflation fell from 6.2% to 5.2% over the same period. However, the level of unemployment has slightly increased from 10.8% in 1984 to 11.2% of working population in 1985.

There is one other very significant factor affecting demand for food in the EEC and that is dietary change. The rate of growth in consumption of certain food products varies considerably. The consumption of sugar, cereals, potatoes, fruit and wine decreased between 1973/74 and 1982/83 while the consumption of meat, vegetables, butter and milk increased (see *Table 3*).

The general population has become more sensitive to dietary issues through concern about weight control and health hazards associated with eating certain foods. The substitution of certain natural foods by synthetic alternatives has also eroded the market for certain agricultural products, as is the case for sugar.

Set against this pattern of decline in the demand for certain foods, agriculture has seen spectacular growth of production over the last 10 years. The overall rate of growth of agricultural productivity has increased by 2% per annum in recent years, and for certain sectors such as cereals, the rate of growth has been even higher at 2.5−2.8% per annum. These productivity increases are chiefly a result of improved yields (see *Table 4*).

Further increases in yields are expected as the impact of genetic engineering reaches the farm in about 5−10 years time, when annual productivity gains of 5% may not be unusual. By conventional breeding and improvement in production techniques sugar beet yields have already been increasing at the rate of 1.9% per annum and this rate is expected to increase. Furthermore, the plant has become more robust and fluctuations in yields resulting from inclement weather conditions are less common. These crops will be produced from a diminishing area of agricultural land. It has been projected that by 1990, with 33 million tonnes of wheat and 2.5 million tonnes of sugar in surplus (Rexen and Munck, 1984), the Community will have 10 million hectares of land in surplus production. By the year 2000, this may increase to 25 million hectares.

World food production imbalance

On world markets there has been a parallel development albeit for different reasons. Several of the world's major commodity markets are now over-supplied as the rate of growth of production has risen sharply amongst exporters whilst the demand from importers has declined. The rate of growth of consumption among developed importing nations has declined for the same reasons as in the EEC: weak population dynamics, dietary changes and the substitution of agricultural products by synthetic foods.

Until recently, it was upon the rate of growth of consumption in developing countries that exporters pinned their hopes. The fast expanding populations of Latin America, Africa and Asia seemed to indicate secure markets for the future, but the general economic recession has curbed demand from these countries. They have suffered from a lack of purchasing power to meet their food requirements, particularly with the recent fluctuations in the value of the dollar. The world sugar market is a classic illustration of the imbalance caused by surplus upon a residual market. Approximately, 100 million tonnes (Mt) of sugar (raw value) are produced in the world, but only 18% of this is freely traded on the world market. Thus, a relatively small deficit or surplus compared with total production can have a dramatic effect upon world prices. The overall trend has been for a steep rise in the level of sugar production among exporting countries whilst the demand for sugar from importing countries has declined (see *Table 5*).

The effect of this imbalance can be clearly seen from the development of the world sugar price over the last 13 years as shown in *Table 6*.

Notwithstanding the general depression in prices, there have been certain years of sharp price increases, notably 1974 and 1980. This is characteristic of world sugar price movements: long periods of price depression interspersed with short bursts of high prices in response to anticipated sugar shortage resulting from natural disasters, such as crop failures, among the world's major exporters. If one considers the variation between the highest and lowest market price quotations in any one year, the extreme volatility of the world sugar market becomes evident. The main reason for this is the

Table 5. World sugar production balance (Mt, raw value).

Year	Production	Consumption	Final stocks	Stocks % of consumption
1974/75	77.9	75.9	17.7	23.4
1979/80	84.9	90.0	25.4	28.2
1981/82	100.7	92.1	32.8	35.6
1984/85	100.3	98.4	41.2	41.9
1985/86 (P)	97.1	99.8	38.2	38.3

From Sugar Economic Yearbook, F.O.Licht, Ratzeburg (Anon, 1984c).

Table 6. World sugar price fluctuations (US cents/lb, raw sugar).

Year	Average	Lowest	Highest
1973	9.48	8.38	14.03
1974	29.71	12.73	63.76
1975	20.43	12.18	45.55
1976	11.56	7.10	15.65
1977	8.11	6.11	10.81
1978	7.82	6.03	9.30
1979	9.66	7.41	15.95
1980	28.69	14.43	43.10
1981	16.83	10.61	31.87
1982	8.35	5.32	13.65
1983	8.49	5.69	12.57
1984	5.20	3.18	7.30
1985	3.61	2.74	5.53

From International Sugar Organisation (Anon, 1986).

small amount of sugar freely traded on the world market. The residual nature of this market is the inherent cause of its instability.

The majority of the sugar produced is either consumed domestically or traded under long-term agreements at prices preferential to those normally observed on the world market. The world's sugar producers have now fallen victim to this two-tier pricing system because a lower-priced sugar substitute has been developed known as high fructose corn syrup (HFCS), made from maize or wheat using modern enzyme technology. HFCS has succeeded in capturing a substantial share of the market in developed sugar-importing nations, as for example the United States, where it has secured virtually 50% of the sweetener market. Further sugar market erosion has come through the advent of low- or non-calorific sweeteners. Biotechnology has given rise to a new generation of artificial sweeteners such as Aspartame which has been readily seized upon by soft-drink manufacturers to boost their sales of diet drinks. These developments are a major cause of concern for sugar producers world-wide. Attempts to control the fluctuations of world sugar prices by an International Sugar Agreement (ISA) have so far been unsuccessful and this helps to explain why the recent protracted negotiations failed to achieve an ISA with economic clauses. The world sugar market is currently in a Darwinistic phase of 'survival of the fittest' at the regrettable expense of a number of developing countries.

EEC sugar production system

In order to protect growers and consumers from the vagaries of the world sugar market, an EEC sugar market organization was introduced in 1968 in the context of the CAP. Its primary objective was to stabilize the internal sugar market whilst maintaining the necessary guarantees in respect of employment and living standards of the Community's beetgrowers.

The sugar sector was the first to apply national production quotas and to adopt a system of financing which ensured that the disposal of any surplus was paid for jointly by growers and processors and thus at no cost to the overall Community budget.

Table 7. National sugar production quotas, 1986/91 (1000 tonnes).

Country	A quota	B quota
Germany	1990	612
Belgium	680	146
Denmark	328	97
France	2996	806
Spain	960	40
Greece	290	29
Ireland	182	18
Italy	1320	248
Netherlands	690	182
Portugal	64	6
United Kingdom	1040	104
Total EEC 12	10 540	2288

In theory, each Member State is awarded a national production quota on the basis of production references in the previous 5 years. It is then the responsibility of the national Government to distribute this quota between the various sugar companies. Sugar produced under the quota enjoys a Price Guarantee. Within the quota there is a distinction between 'A' and 'B' quota sugars. 'A' quota sugar makes up the majority of the quota, between 70 and 90%, and receives 98% of the Guarantee Price. 'B' quota receives between 60.5 and 98% of the Guarantee Price — the differential in each case being accounted for by the sugar beet production levies. The level of national quotas are reviewed every 5 years and the result of the most recent review of the 1986/91 period is given in *Table 7*.

Sugar produced outside the quota is termed 'C' sugar and is not eligible for Price Guarantee on the international market. It has to be sold on the world market without export subsidy for whatever price it will fetch.

The Sugar Price Guarantee system consists of three parts:

(i)   the *Target Price* representing the price which might be expected to apply in the market place in a situation of balanced supply and demand;

(ii) the *Threshold Price* which represents the minimum selling price for sugar imported from third countries;

(iii) the *Intervention Price* which is the price at which excess sugar may in principle be sold into intervention — in practice intervention selling is little used.

In 1986/87 the EEC Institutional sugar prices were:

(i)   white sugar Threshold Price   66.86 ECU/100 kg;

(ii) white sugar Target Price   57.03 ECU/100 kg;

(iii) white sugar Intervention Price   54.18 ECU/100 kg.

From the Intervention Price, the support price for sugar beet is derived using a theoretical yield calculation of 130 kg of white sugar from 1 tonne of sugar beet at 16% content. There is an institutionally fixed sugar processing margin which for 1986/87 was as follows:

(i)   Revenue: Intervention Price of white sugar, 54.18 ECU/100 kg + receipts from

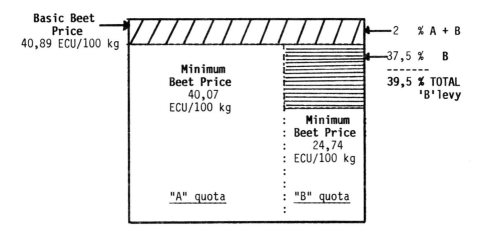

Figure 1. EEC sugar production levies 1986/87.

sales of molasses, 1.89 ECU/100 kg = Total receipts, 56.07 ECU/100 kg.

(ii) Expenses: Value of beet, 31.45 ECU/100 kg + cost of transport, 3.73 ECU/100 kg = Total cost, 35.18 ECU/100 kg. Total processing margin, 20.89.

In practice, the yield of sugar from sugar beet is in fact higher than 130 kg. The Guarantee Price for sugar beet is known as the *Basic Beet Price*.

The system of sugar production levies has recently been modified. In principle, it consists of two parts: a basic production levy of 2% on the Basic Beet Price which applies to all sugar produced under the quota, both A and B. In addition, there is a variable levy upon 'B' quota sugar which may be applied up to a maximum level of 37.5% of the Basic Beet Price (in addition to the basic levy of 2%). The level of this production levy is fixed annually in accordance with the cost of exporting the surplus of quota sugar onto the world market. This system is explained diagrammatically in *Figure 1*.

With the recent world market price depression, the maximum levies applied to quota sugar were inadequate to meet the costs of disposing of surplus quota sugar on the world market. Hence an additional levy was imposed on quota production representing on average 1.3% of the Intervention Price of white sugar although this amount was differentiated between Member States according to their level of contribution to the costs of the Sugar Regime. This additional levy will apply for the duration of the 1986/91 Regime and is designed to recover the debts accumulated from the previous 1981/86 Regime.

The heavy burden of meeting the costs of Community sugar exports has led to a decline in the area of sugar beet sown in the European Community since 1981/82 (see *Table 8*).

Some of the impact of reducing beet area has been mitigated by improvements in yield. This is most evident from 1985/86. Over this same time period domestic sugar consumption has remained virtually static at 9.5 Mt. Not only have the quantitites of quota sugar for export been reduced, but growers producing 'C' sugar beet, who are exposed to the full impact of the world price at the margin, have been severely cutting

Table 8. EEC sugar production trends 1981/82 – 1985/86.

Year	Area 1000 hectare	Yield tonnes/hectare	Production
1981/82	2022	7.26	14.71
1982/83	1838	7.41	13.63
1983/84	1667	6.42	10.71
1984/85	1727	6.94	11.96
1985/86	1719	7.23	12.43

From European Committee of Sugar Manufacturers (CEFS).

back. In France, a large 'C' sugar producer, reductions in beet area have been announced between 8 and 10% for 1986/87 and this is chiefly a 'C' beet area. Many factories are now running well under capacity — the largest French factory Conantre, slicing 25 000 tonnes at only 50% — and some will be forced to close. The same situation applies in Belgium and West Germany.

The search for new outlets

Drawing together the strands of this argument, it is clear that a number of factors have combined to make the search for new outlets imperative: ever-increasing productivity, stagnant or declining levels of domestic consumption and depressed export markets. The full measure of this impasse confronting arable farmers became the subject of a major reflection on the need to reform the CAP by the European Commission published under the name of a 'Green Paper' in July 1985. Com. (85) 750.

The purpose of this document is to indicate a number of alternative options to extract agriculture from this predicament. These are:

(i)   reform of sectors in surplus production, notably cereals;
(ii)   exploration of alternative productions, proteins and oil crops;
(iii)   development of industrial and energy uses of agricultural raw materials.

This is not the place to enter into an exhaustive commentary on the Commission's proposals for the reform of the Cereals Regime, nor on their intentions to promote alternative crop productions, but rather in accordance with the title of this chapter to examine the potential of new non-food uses of arable crops.

Non-food uses of agricultural raw materials are not a novelty. For a long-time, it has been possible to supply sugar and starch to the fermentation industry for the manufacture of a certain number of listed products. The present uses of these substrates has been limited by the Community pricing policy which rendered these raw materials unattractive compared with those in third countries. If this situation were changed, the European Council of Chemical Manufacturers' Federations (CEFIC) has estimated that consumption of carbohydrates might increase from 1.7 Mt to 3.1 Mt in the short term and to 4.6 Mt in the long term (see *Table 9*).

Vegetable fats and oils are an important raw material for the European Community chemical industry. At present, a considerable proportion of these have to be imported as suitable plants are not available in the EEC. Historically, plants have been bred to yield oils and fats for human consumption, but with different breeding criteria it will

Table 9. Use of renewable resources by the chemical industry in the EEC (1000 tonnes).

	Present use	*Estimated use*
Sugar	100	500
Starch	400	1000–2450[a]
Starch (in other industrial sectors)	860	1600
Total starch	1260	2600
Oils and fats	1700	3500

From CEFIC.
[a]Long term potential

be possible to produce substitutes for imported oils. This possibility creates interesting opportunities for arable farmers looking to reduce the area of their land under surpluses.

The consumption of sugar by the EEC chemical industry has been declining on account of the non-competitive price of EEC sugar compared with that obtainable on the world market. Production refunds were available on sugar for certain listed products but these were inadequate to cover the difference between the EEC price of sugar and the price at which competitors could obtain their sugar. To be precise, the refund available only covered 10% of this differential. This discrepancy led to the absurd situation where EEC chemical companies were transferring production outside the EEC in order to obtain cheaper sugar – often EEC sugar exported onto the world market with a restitution covering the full difference paid by the grower. A good illustration of this is the purchase of a West German citric acid manufacturing company, Boehringer Ingelheim by an Austrian company, thus making Austria the world's largest producer of citric acid. By this means, European farmers have actually been paying to deprive themselves of a less costly domestic outlet.

Clearly such anomalies need to be rectified and the good news is that a reform of the regulations governing the sale of sugar and starch to the chemical industry was achieved in April 1986. It has been agreed that from August 1, 1986, sugar and starch will be available at competitive prices, using refunds based on the difference between EEC and world prices.

In this reform, it was most important to achieve equilibrium between sugar and starch, since for many years there has been a distortion of competition in favour of starch, with sugar losing considerable market share to the starch sweetener, glucose. In the new Regulation 1006/86, it is categorically stated that for the first two campaigns 1986/87 and 1987/88 the price of sugar will be equal to the price of glucose. Thereafter, there may be adjustment towards the world price of sugar, but, in any event, the effect of these regulations on the development of these new markets will be reviewed in 1989/90.

There are two further strong arguments in favour of maintaining equilibrium between sugar and starch.

(i)   One is that they are completely technically interchangeable and need to be put at the disposal of the chemical industry on equal terms.

(ii) The second is that since the arable farmer produces both raw materials for sugar and starch, it is not in his interest to see a price-cutting war between competing crops.

Table 10. Forecast sugar up-take by the EEC chemical industry (tonnes).

	1978/79	1984/85	1994/95
Organic acids	11 057	8790	168 000
Organic alcohols	12 741	8994	–
Fructose	5670	10 861	14 000
Penicillin	25 804	14 885	40 000
Other antibiotics	5591	665	8000
Adhesives	1023	695	4000
Other conventional chemicals	20 386	17 283	100 000
New biodegradables	–	–	200 000
Total	82 272	62 092	534 000

From CEFIC.

The production refund level will be calculated periodically and may be set for the duration of a complete marketing year. The logic behind this is that the chemical industry does not require sugar at spot on world price, which is liable to daily fluctuations, but rather at a stable price ensuring security of supply and quality, in order for an investment decision to be made.

The arable farmer is interested to know what his crops will be used for by the chemical industry. He is familiar with the traditional chemicals for which his starch and sugar have been used in the past, such as organic acids and alcohols. But there is little room for expansion of these markets worldwide and it is more a question of recovering markets lost to third countries when EEC raw materials were non-competitively priced. Of greater interest is the new area opening up as a result of the application of biotechnology. It is in the category of speciality chemicals where real growth is expected to occur (see *Table 10*). These high value chemicals do not require large volumes of agricultural substrate, but nonetheless constitute an interesting new opening.

Awareness of the changing political climate in favour of substitution by environmentally preferable products has stimulated research into biodegradable substitutes, such as detergents and plastics by EEC chemical companies. In order to demonstrate the magnitude of these outlets, an example has been supplied by Imperial Chemical Industries (ICI) p.l.c. for the manufacture of biodegradable plastic [polyhydroxybutyrate (PHB)].

ICI have developed a process whereby a microorganism produces fat from a carbohydrate substrate from which plastic can be manufactured capable of substituting existing polypropylene plastic with the additional property of biodegradability in film form. At present, 1 Mt of oil-derived, non-degradable plastic are consumed annually in the Community. At least some of this market could be captured on condition that the carbohydrate substrate is supplied at a competitive price level, although the evolution of oil prices is a determining factor in this assumption. Conventional plastic prices range from 2500 to 3000 ECU/tonne in the Community. In order to compete in the lower price range of oil-derived plastics, the carbohydrate substrate would need to be supplied at 320 ECU/tonne which corresponds to 60% of the Intervention Price of white sugar. Within the present budgetary estimates for cereals under the Starch Regime, it would

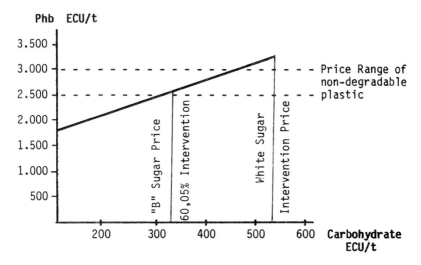

Figure 2. Cost calculation for biodegradable plastic.

be possible to provide a restitution on starch enabling glucose to be supplied at the same price. The cost calculation for biodegradable plastic is shown in *Figure 2*.

Research has shown that the optimum plant size is for the capacity of 30 000 tonnes of plastic per annum, requiring 120 000 tonnes of carbohydrate substrate, entailing an investment of £40 million, creating 50−70 jobs in research, production and sales and located preferably in a rural area adjacent to the source of raw material. It is also possible to manufacture biodegradable, flame-resistant plastic from starch by chemical processes. Recently, the Italian government passed legislation prohibiting the use of non-degradable plastic wrappings by 1993. So the stage is set for PHB.

Although an interesting and encouraging outlet for the European arable farmers, the total projected use of carbohydrates for these purposes by the early 1990s of 3.1 Mt will only make a small dent in the anticipated surplus of 33 Mt of cereals and 2.5 Mt of sugar expected to be available. This is what has prompted arable farmers to seek an outlet of greater proportions. Just such an opportunity has presented itself with the decision to ban lead from petrol in the European Community by 1989.

The removal of lead from petrol has the effect of lowering the octane content, thus requiring engines to be modified or octane boosters to be added to restore octane levels. Ethanol of agricultural origin can be successfully used as an octane booster and when added to petrol in the proportion of 5% can restore two octane points lost by the removal of lead. If 5% bioethanol were added to petrol, there would be a potential market for 11 Mt of cereals and 2.5 Mt of sugar if these raw materials were used in a ratio of 2:1.

The logic for using this beet/wheat mix is to combine the advantages of sugar beet, which has a higher yield of ethanol per hectare, with those of wheat which has a high value protein by-product, vital wheat gluten, as well as the ability to be stored for longer periods than beet thus permitting ethanol plants to function all year round. A further reason is provided by the need to maintain a balance between cereal and root crops in the arable rotation. Beet and wheat are by no means the only raw materials suitable

Table 11. Agricultural raw materials: ethanol yield per tonne of raw material and per hectare of crop.

Raw material	Yield (tonnes per hectare)[a]	Alcohol yield per tonnes of raw material [hectolitres (hl)]	Alcohol yield per hectare (in hl)
Beet	53.5	1	53.5
Wheat	5.5	3.6	20
Maize	6.6	3.6	24
Barley	4.6	3.1	14
Potatoes	29.5	1.053	31
Wine	75.1	0.1	7.5
Roasted chicory	36.8	–	–
Sorghum	4.7	3.3	15
Jerusalem artichoke	55[b]	0.85[b]	47
	66[c]	0.90[c]	59
Carob-beans	5.6	1.75	10

[a]EEC of ten average.
[b]According to 'Agricultural Information' No. 564, January 1985.
[c]According to a study from Lille's Agriculture Institution, France.
From the European Commission.

Table 12. Cost of producing ethanol (in ECUs) from different raw materials (based on production of 1 hl of alcohol)[a].

	Wine	Potato (1)	(2)	'A' beet (1)	(2)	'B' beet (1)	(2)	Maize (1)	(2)	Wheat (1)	(2)
Raw material cost	342	57	45	40.07		24.74		68.1		53.44	
Transport		3	3	4		4					
Processing cost	46	30		25	20	25	20	30	20	30	20
Value of by-products		5	5					20		20	
Total (ECU/hl)	388	85	73	69	64	54	49	78	68	63	53

[a]The columns represented by the numbers (1) and (2) give the estimated processing costs for beef, wheat and maize, according to whether the specialized production of ethanol is being carried out in small- (1) or large-scale plants (2).
From European Commission, Rapport de Synthèse Bioethanol (1985).

for ethanol production. Ethanol can be obtained from a host of other arable crops which may suit particular regional circumstances, as is the case for Jerusalem artichoke and sweet sorghum in Mediterranean areas (see *Table 11*).

The objective must be to obtain an optimum crop mix for the particular locality.

There is a huge variation in the cost of bioethanol from the various agricultural sources. The cost of the raw material is basically dependent upon yield. The processing costs are undoubtedly the greatest variable.

Until a genuine large-scale production operation has been carried out on Community territory, it will be necessary to rely on the costs resulting from experiments carried out on a small-scale or in third countries. The European Commission acknowledges that, once specialist ethanol production plants have been built, in which large economies of scale will be achieved, the lower of the two processing costs given in *Table 12* will be the relevant figure.

Table 13. Price of bioethanol and its competitors (ECU/hl).

	Market price	*Production cost*
Agricultural ethanol	20−35	49−63
Methanol	15−16	13−15
TBA[a]	38−36	29−32
MTBE[b]	47−42	28−33

[a]Tertio-butyl acid.
[b]Methyl-tertio-butyl ether.
From European Commission Rapport de Synthèse Bioethanol (1985).

As the processing operation is very similar, the costs are all very much alike. The variable value of the by-products also has a strong impact on the overall cost of bioethanol from the various sources.

The combination of raw material and transformation costs, together with by-product credits, gives a remarkably similar production cost for the majority of agricultural raw materials.

The proponents of bioethanol have encountered much criticism and obstruction, chiefly orchestrated by the oil industry, which is unwilling to have to depend on agriculture for a part of its production process. Objections tend to be raised that ethanol costs more to produce than its competitors of fossil fuel origin. This is true for methanol which is currently the cheapest form of oxygenated compound derived from oil. Yet it is conveniently forgotten that methanol will not readily mix with petrol and requires a more expensive co-solvent. These co-solvents of oil origin are more expensive and not yet available in the Community in adequate quantities. Bioethanol can, in fact, serve perfectly well as a co-solvent. The relative cost of these oxygenated compounds can be seen in *Table 13*.

It will be immediately apparent from *Table 13* that bioethanol costs more to produce than its current value in the market place. This is a result of the unrelated development in the price of agricultural raw materials and the price of oil.

Economically, the case for ethanol is usually presented on too narrow a basis. There is a failure to take account of the additional octane points which ethanol brings to petrol which, if expressed in usage value, would be 1.2 times that of petrol. Furthermore, for most Member States in the Community, petrol is imported, requiring an outlay of foreign exchange. If the savings on imports and the generation of wealth from an alternative domestic product are taken into account, bioethanol more than pays for itself. It is on these grounds that the agricultural lobby claims that the difference in price between ethanol and petrol should be met by de-taxation, national governments taking account of the macro-economic benefits which bioethanol would bring.

There is a further important macro-economic argument in favour of bioethanol at Community and one very dear to the arable farmer's heart. This concerns the use of the Common Agricultural budget.

There is a recent development towards making farmers financially responsible for the cost of disposing of their production surpluses. This has recently been achieved by the introduction of a co-responsibility levy on cereals and such a system of self-financing has long been in place for sugar. Thus, the arable farmer must ask why the same amount of money used to dispose of his surplus on world markets should not rather be used to invest in an alternative domestic outlet and at the same time bring relief to depressed world markets? This thought was given expression by the Agricultural Commissioner at a recent meeting of the French Committee for International Agricultural Relations, Commissioner Andriessen made the following announcement on bioethanol: 'As far as bioethanol is concerned, already in our Green Paper and then in our Memorandum on cereals, we envisaged the possibility of providing an aid equivalent to the export restitution, both for cereals and for sugar beet transformed into bioethanol'.

However, there is no mention of when this re-distribution of agricultural funds might take place.

The critics of bioethanol also invariably produce arguments on technical and energy grounds but these can be convincingly dismissed. The oil industry has argued that the energy balance for bioethanol is negative, that is, more energy has to be put into ethanol production than is ultimately gleaned. Pilot plants in France and Germany have proved that with energy-saving techniques, such as the generation and use of biogas from the by-products of ethanol production, the energy balance is invariably positive and may be as high as 1:2, that is, one unit of energy consumed for two units produced. This has been achieved by energy saving in the fermentation process and re-cycling of waste for biogas. A further technical criticism is the suitability of ethanol as a petrol additive, but the fact that, in the US, cars run on a 10% ethanol admixture without difficulty should dispel these fears.

These two new non-food outlets are very different in scale and feasibility for agriculture. The opening of the enlarged outlet to the chemical industry was relatively easy to achieve. The opening of the outlet for bioethanol looks like being more difficult. Both have important implication for European farming.

Concluding remarks

What we are witnessing is the transformation of the CAP from an exclusively food policy with the accent on security of supply, to a combined food and non-food policy with the accent on competitiveness and versatility. This represents an enormous challenge to European agriculture, so diverse in its nature.

Those of us engineering this agricultural metamorphosis need to take care not to destroy the positive aspects of the existing situation in the process. The new non-food outlets will only gradually expand and it is thus important to maintain a soundly economic food outlet in order to allow these new openings to grow. The CAP may be in need of reform but it is not all bad. On the contrary, many of its principles still hold good today and will be needed just as much in the future, notably the principle of Community Preference and the need to prevent distortions of competition between agricultural products.

Where more thought needs to be given is to the developing relationship between agriculture and the non-food industries, or more specifically between farmers and chemists, biologists and genetic engineers. Fruitful collaboration will yield solutions to the agricultural imbalance by finding an ever-increasing number of renewable alternatives to our presently wasteful consumption of the world's finite resources. But both sides need imagination and understanding for the technical and economic constraints limiting one another's potential. I sincerely hope this report goes some way towards exposing agriculture's great potential and its attendant problems in order that together we find solutions in the interest of our industries, our economy and society as a whole.

References

Anon. (1984a) *The Agricultural Situation in the Community*. Report 1984, Commission of the European Communities, Official Publications Office, Luxembourg.

Anon. (1984b) *Eurostat Review 1973/82*. Statistical Office of the European Commission, Luxembourg.

Anon. (1984c) *International Sugar Economic Yearbook and Directory*. F.O.Licht, Ratzeburg, FRG.

Anon. (1985a) *Basic Statistics of the Community*. 23rd Edition, Statistical Office of the European Commission, Luxembourg.

Anon. (1985b) *The Agricultural Situation in the Community*. Report 1985, Commission of the European Communities, Official Publications Office, Luxembourg.

Anon. (1985c) *Zuckerwirtschaft 1985/86*. Verlag Albert Bartens, Berlin.

Anon. (1985d) *Bulletin Statistiques 1985/86*. Fonds d'Intervention et de Régularisation du Marchè du Sucre, Paris.

Anon. (1986) *Statistical Bulletin*. Vol. 45, No. 4, International Sugar Organisation, London.

Harris,S., Swinbank,A. and Wilkinson,G. (1983) *The Food and Farm Policies of the European Community*. Wiley Press, Chichester.

Munck,L. and Rexen,F. (1984) *Cereals Crops for Industrial Use in Europe*. Commission of the European Communities and Carlsberg Research Institute, Denmark.

CHAPTER 3

Carbohydrate feedstocks: availability and utilization of molasses and whey

J.COOMBS

Bioservices, London and Biotechnology Affiliates, Reading, UK

Introduction

Sucrose is widely used as a fermentation substrate for the production of high volume/low value products such as ethanol, single cell protein (SCP), organic and amino acids and some microbial gums. It may also be used to produce higher value/low bulk products such as antibiotics, speciality enzymes, vaccines and fine chemicals. However, particularly in the case of the bulk products, it is seldom used in the pure form. Rather, a number of by-products of sugar processing (generally known as molasses) are used. (For simplification and in order to use terminology familiar to the industry the term 'sugar' is used to denote sucrose. All other sugars are referred to using their common chemical name.)

Historically the single objective of most sugar factories and refineries was to produce a high quality sugar leaving as little as possible in molasses which was, and in many cases still is, regarded as a waste material of little value. However, an increased interest in the production of fuel alcohol in particular, as well as a position of sugar surplus, has changed this situation in some regions. Where large fuel alcohol installations have been built, such as in Brazil, the clarified juice stream may be taken direct to the fermenter. Other alternatives which have been adopted include the direct fermentation of molasses within the sugar factory complex.

Recognition of the fact that sugar processing establishments can become multiple-product factories has led in some cases to suggestions for modification of the product streams in such a way that less sugar is extracted from the process, resulting in a purer syrup which may be used as substate in-house or may be marketed. The production of such fermentation syrups is not a new development however, since traditionally concentrated (inverted) cane juice has been marketed as the so-called 'high test' molasses.

The type of fermentation which has been developed to use molasses is often quite robust and able to accept other low-purity carbohydrate wastes or by-products. Since these find similar uses, and may supplement or compete with molasses and other syrups produced by the sugar industry, some aspects of products such as paper-pulping liquors, citrus fruit canning wastes (citrus molasses) and cheese whey will also be considered.

Feedstock availability and costs are considered here mainly from a European viewpoint. However, since there is considerable interaction between the European beet-based sugar industry and the cane sugar industry through the Lome Convention (1986) and the EEC is a net importer of molasses, both the cane and the beet sugar industries are considered in terms of agriculture, processing and price of products on the basis of fermentable sugar available to the consumer.

The major impact on costs within Europe, as compared with overseas markets, is from EEC legislation. Since this is dealt with in detail in Chapter 1, the emphasis here is on a comparison of the price paid for their raw materials by overseas manufacturers as compared with the price in Europe, which at present limits the expansion of many of the areas of fermentation where feedstock costs represent a significant proportion of the final selling price. This is true for fuel alcohol production where raw material costs are from 60 to 75% of the pre-tax selling price.

Crops and sugar markets

The two main sources of sucrose are beet (*Beta vulgaris*) in the temperate regions and cane (*Saccharum officinarum*) in sub-tropical and tropical regions. The sugar in beet is stored in the enlarged tap-root and hypocotyl whereas in cane the sugar is stored in the stem. As far as Europe is concerned most of the sucrose produced comes from beet with some cane grown in the South of France and Spain. The major cane sugar producers are Brazil, Cuba and India with significant quantities from the Far East, Australia, South Africa, Latin America and the USA (mainland, Hawaii and Puerto Rico). Beet is an annual crop planted as seed and harvested in the autumn. In contrast cane is perennial, although the duration of the growth period, frequency of harvest and number of crops taken before re-planting from stem cuttings varies greatly from one region to another.

As with other European crops there has been a steady growth of 2−3% per annum in sugar production reflecting genetic improvements, increased chemical inputs and higher degree of mechanization. The main genetic advance in Europe has been the development of triploid varieties producing monogerm seed (Campbell, 1976) (that is fruits which give rise to single plants rather than clusters) thus avoiding the expense of thinning the seedlings. The emphasis in breeding has been for increased yield of high purity juice which will give a good recovery of crystallized sugar, rather than total carbohydrate or fermentable sugar. Other characteristics which are favoured relate to disease resistance (against aphids and virus disease), root size and shape to favour mechanical harvesting.

Although average yields of beet have risen in the UK from around 30 tonnes per hectare in 1973 to over 40 now (CSO, 1985) these yields are lower than those found in the other major EEC producing countries (France and Germany). At the same time quotas restrict the total amount of A and B sugar (see Chapter 2) which can be produced; hence as yields increase planted area decreases (from ~210 000 hectares in 1981 to <200 000 in 1984/5).

Before the entry of the UK into the EEC about two thirds of the domestic needs for sucrose were filled by imported cane sugar. Following entry the commitment to certain developing countries in the ACP (African-Caribbean-Pacific) countries was incorporated into the Lome Convention (1986) with a preferential import of 1.3 million tonnes (Mt) at a guaranteed price. Again there has been an improvement in yields due to a combination of breeding (for juice purity, sugar yield and disease resistance) and technical inputs. But in general a combination of political instability, depressed sugar price and labour problems have resulted in decreased cane sugar production in many of the traditional areas of cultivation. The major exception to this general trend is that associated with the rapid expansion of the Brazilian fuel alcohol programme which now has installed capacity to produce over 12 Mt of hydrous ethanol per annum.

Processing

The characteristics of the various sucrose-containing raw materials used for fermentation substrates depend on the nature of the crop used, the extraction and purification technology adopted and the product or by-product stream selected for use. Hence these aspects are now considered in brief.

The processing of both cane and beet is similar in that sucrose is recovered by crystallization from a thick syrup derived by concentration under vacuum of a clarified juice. The crystals of sugar are then removed by centrifugation and the liquid residue recycled through two or three further crystallization stages until the concentration of impurities in the residual syrup is such that no more sucrose can be economically recovered. This residual syrup, generally known as molasses, is the most widely used fermentation substrate derived from the sugar industry.

Although the objective with both crops is to optimize the yield of white crystalline sugar, the processing of sugar cane and beet differs. White sugar is generally produced from beet in a single factory operation in the country of origin whereas the production of white cane sugar is a two-stage process. Traditionally impure (raw) cane sugar was produced in the various colonies of the developed countries, shipped to Europe or the USA and then refined. Nowadays more white cane sugar is produced in the countries of origin for own consumption − but the need for two stages of processing remains.

Beet is mechanically harvested and the de-topped roots are transported, generally by road to a nearby factory, where they are weighed, sampled for sugar content, washed and sliced into 'cossettes'. These are then fed into a counter-current diffuser which leaches out the sugar and the resultant juice is treated with lime and carbon dioxide to remove impurities. The precipitate of calcium carbonate is removed and the juice concentrated, decolourized using carbon or resins and the sugar recovered by several stages of crystallization. The various impurities become concentrated in the syrup stream which finally forms the beet molasses. Beet differs from cane in that it contains up to 2% of solids as raffinose (D-galactosyl 1,6-D-glucosyl 1,2-D-fructoside). This trisaccharide, which is a crystallization inhibitor, accumulates in the molasses. The residual pulp is normally dried, mixed with molasses and used as animal feed. The process requires an outside source of fuel.

Cane may be harvested mechanically from large commercial plantations, although large quantities are hand cut from small-holdings. Transport is equally varied ranging from animal power, through road and rail to boat. As with beet there is a need to get the material to the factory with as little delay as possible in order to reduce biodeterioration. Cane that is left in the field, transporter or factory holding bay for too long may become infected with the dextran producer *Leuconostoc mesenteroides* in particular. Dextran and other polysaccharides can cause problems in crystallization, contribute to the formation of haze (acid beverage floc) when used in carbonated drinks and may accumulate in molasses. Stored cane may also contain an increased proportion of glucose and fructose due to enzymic hydrolysis of sucrose (inversion). The proportion of carbohydrate recovered as molasses is also increased where the juice contains high levels of potassium, which is taken up by the growing plant and thus depends on the soil type.

At the factory the cane is cleaned of trash, washed and passed through shredding knives and squeezed through rollers. The expressed juice is heated and limed to precipitate impurities which are removed by filtration or flotation. Various processes (carbonation, sulphitation or phosphatation) are used in which carbon dioxide, sulphur dioxide or phosphoric acid and polyacrylamide are added to aid precipitation or flocculation. Residues of these will be found in the molasses which will differ according to the process used. The clarified juice is concentrated by evaporation and boiled in steam-heated pans under vacuum to produce a mixture of about 50:50 crystallized sugar and thick juice from which the raw sugar is recovered by centrifugation. The syrup will pass through further stages of crystallization until further sugar cannot be economically removed leaving factory as blackstrap molasses. The cane waste (bagasse) is used to fuel the factories making the process self-sufficient in energy.

Raw sugar is traded through preferential agreements or enters the world market before being further processed, usually in a developed country, in a sugar refinery. The raw sugar is screened for metal and other impurities and the outer layers (which contain most of the impurities since they are formed from dried-on molasses) softened with warm syrup, the crystals washed with water and recovered by centrifugation. The syrup and washings will go to a recovery house, while the crystals (recovered by centrifugation) are dissolved in water, treated with lime and carbon dioxide, filtered and the syrup decolourized using carbon. The now colourless liquid streams then goes to vacuum crystallizers and the sugar is recovered by centrifugation and dried in a rotary air drier. As with raw sugar several stages of crystallization occur. The residual syrup with a high impurity content then goes to the recovery house where further sugar is extracted finally leaving factory molasses. It is not unusual for some beet sugar to be put through what is essentially a cane refinery. This can cause problems in the recovery house since the inhibitory effect of raffinose is to block development of the end faces of the crystal resulting in squarer crystals. In contrast dextran (found in raw cane sugar) inhibits growth on the long faces of the crystal causing needle shapes which are more easily broken during centrifugation. A combination of raffinose and dextran can block crystallization completely, resulting in poor recovery and overproduction of molasses.

Fermentation substrates from the food industries

Sucrose based products

The main products from the sugar industries are as follows:

(i)   refined white sugar of over 99.7% purity, derived from either sugar cane or sugar beet;

(ii)   raw cane sugar which by definition contains less than 99.5% by weight of sucrose in the dry state;

(iii)   molasses, of which several distinct types may be identified as blackstrap (cane factory) molasses, refinery molasses, beet molasses or high test molasses;

(iv)   intermediate process streams;

(v)   cane juice or beet juice.

Table 1. Sugar production, trade and stocks 1984/5 (million tonnes) (FAO, 1985).

	Production	*Consumption*	*Imports*	*Exports*	*Stocks*
World	97.55	96.14	–	–	41.6
EEC	13.3	9.4	1.45	5.07	3.8
UK	1.3	2.3	1.1	–	0.8

Traditionally molasses was the main substrate used in sucrose-based fermentation. However, once molasses started being used on the production site, in annexed distilleries for example, it became logical to increase fermentation capacity by using the fresh juice, or taking one of the clarified and partially concentrated intermediate streams. Under these conditions the site changes from being solely a sugar factory to a multi-product system which can produce a wide range of food, animal feed and chemical products as detailed below.

The degree of purity of both refined sugar and most raw sugars is such that any impurity is likely to have little effect on a fermentation, either due to the presence of inhibitory substances or the effects of potential nutrients. However, the reverse is true of molasses, some of the intermediate product streams or direct expressed juice. Hence, the physical characteristics and chemical composition of these materials will be considered in more detail. However, since molasses is a by-product of the sugar industry the availability reflects the size and characteristics of the white sugar market, *Table 1*. Over the last 5 years the total global production of sugar has varied between 90 and 101 Mt per annum whilst consumption has varied between 90 and 96 Mt. Europe as a whole produces about 13 Mt of white sugar, importing in addition about 1.5 Mt of cane sugar (much of which comes into the UK) under the Lome Convention (1986) and exporting surplus production at a level which has ranged between 3 and 5 Mt per annum over the last few years.

Molasses

Approximately 300−360 kg of molasses are formed for each tonne of sugar produced (Parker, 1982), hence although detailed statistics are not kept, it may be assumed that some 30−40 Mt of molasses are potentially available. In reality some may not be readily available at an economic price due to the isolated position and lack of transport facilities associated with many cane factories. The EEC produces about 3.5 Mt of molasses (out of some 10 Mt produced in Europe as a whole) and imports about 3 Mt. Much of this is used as an animal feed supplement (Baker, 1982).

Process streams and juice

There are no practical problems in removing any of the liquid streams at any stage of either raw or white sugar production from cane or beet. At present clarified cane juice is fermented directly to ethanol in some 200 distilleries in Brazil and in Europe (in France for example). The first extract from beets may be used on a much smaller scale in the same way. The benefits of taking a single strike (single crystallization) in either a cane or beet factory and then using the residual liquor for the production of fuel alcohol is technically attractive where sugar surpluses exist. The properties of such syrups would fall between those of molasses, where the optimum amount of sugar has been removed, and high test molasses where no sugar has been removed, although partial or total inversion is carried out in order to prevent crystallization during transport. There are advantages in using intermediate process syrups or high test molasses rather than final molasses (Alexander, 1985). These advantages include:

(i)   higher volumetric sugar content;
(ii) lower concentration of impurities;
(iii) lower storage costs;
(iv) greater chemical stability;
(v)   fewer problems with waste or stillage disposal.

The quality of the fermentation may also be improved as there is less risk of obtaining a non-fermentable batch of feedstock as may happen with final molasses.

Obviously in an integrated factory system there are advantages in going one step further and during the processing campaign using clarified juice directly. The advantages of doing this include:

(i)   lower process costs;
(ii) the need for less purification;
(iii) a reduced energy requirement for substrate concentration;
(iv) no equipment, holding tanks or clean water needed for substrate dilution;
(v)   no liquid storage costs;
(vi) an overall decrease in capital cost.

However, if the process is restricted only to juice use there will be both seasonal and geographical constraints on production. Hence, logic may dictate a combination of the use of juice in season with storage of thick juice or use of bought-in molasses for the rest of the year.

Whey

Whey is a by-product of the cheese-making industry that arises following the separation of curds (the solidified casein and butter fat). As detailed below the solids content of fresh whey is very low which makes its use problematical if, as is the case in many regions, it is only available from a large number of scattered outlets each producing limited volumes. The fact that dilute whey is a good substrate for the growth of many organisms causes further problems since some sort of preservation is required during transportation. Thus, the material can either have a value or become a source of pollution and hence an expensive waste for smaller units. On the other hand for larger units, with an output in the range of $0.5-1.5$ million litres per day, economic uses can be found and the whey processed in a number of ways to make it easier to handle, transport and store. Products include:

(i)   *whey molasses*, a viscous liquid produced by concentrating whey, to around 70% solids, often using reverse osmosis followed by evaporation;
(ii) *whey permeate*, a liquid containing about $35-50$ g/l of lactose produced as a by-product of whey protein recovery;
(iii) *whey powder*, a dry power produced by concentrating whey to 50% solids by evaporation and then spray drying.

Table 2. The composition of molasses from various sources (Baker, 1982; Paturau, 1982).

Carbohydrates	Percentage values of				
	Solids	Sugars	Sucrose	Invert	Raffinose
Blackstrap	80–86	50–65	30–40	10–25	
Beet	76–85	48–58	47–55	0.2–2	0.2–2
Refinery	76–84	50–58	32–42	14–20	
High test	82–86	72–75	tr	72–75	
(Non-sugar organic matter 9–20%, ash 5–12%)					

Vitamins (mg/kg)	Cane	Beet
Biotin	3	0.4
Folic acid	0.04	0.2
Inositol	6000	8000
Pantothenate	55	100
Pyridoxine	3	5
Riboflavin	3	0.4
Thiamine	2	1.3
Nicotinic acid	800	45
Choline	600	400

Minerals (%)	Cane	Beet
Sodium	0.1 –0.4	0.3 –0.7
Potassium	1.5 –5.0	2.0 –7.0
Calcium	0.4 –0.8	0.1 –0.5
Chloride	0.7 –3.0	0.5 –1.5
Phosphorus	0.03–0.1	0.02–0.1
Sulphur	0.3 –0.8	0.15–0.5

The whey may be partly or totally demineralized by electrodialysis or ion-exchange, deproteinized by ultrafiltration, or both demineralized and deproteinized.

Chemical composition of fermentation substrates

Molasses

As shown in *Table 2* the major carbohydrate components of molasses produced by the sugar industry are sucrose, glucose and fructose. Beet molasses characteristically contains only small amounts of glucose and fructose and additionally contains significant levels of the trisaccharide raffinose. All molasses has a high inorganic ash content and contains a number of different vitamins and growth factors as well as nitrogenous and other organic compounds which include chlorogenic and caffeic acids, uronic acids, sugar alcohols, organic acids, amino acids, nucleotides, sterols, tannins, plant pigments, gums, waxes and lipids. In particular beet molasses has a high concentration of non-amino acid nitrogen-containing substances such as betaine and various polyamines.

Other similar wastes which may be admixed with sugar molasses include concentrated waste streams from citric processing with similar sugars, from maize-based starch syrups (hydrol) with a high glucose content, and pulping liquors with a high pentose (xylose) content. When added these may be detected by analysis revealing these additional sugars or characteristic phenolic compounds or lignin sulphonates.

Molasses is not a sterile material. Problems may arise from the presence of micro-organisms and osmophilic yeasts, which will grow under conditions of up to 65% dry matter, in particular. Molasses may also contain spores of fungi and bacteria or resting cells which are prevented from growing in the bulk material due to osmotic effects but may grow in storage tanks at the liquid air interface where condensation results in a lower dry matter concentration.

Whey

The exact composition of whey differs according to the animal from which the milk was obtained and the type of enzyme (rennet) or lactic fermentation used for a particular cheese (*Table 3*). The coagulation with rennet yields sweet whey with a high lipid content. Coagulation by lactic fermentation yields acid whey, containing smaller quantities of lactose and proteins. The solids content of whey is around 6−8% with lactose the major component (4−5%). This lactose may be 70−80% of the total dry solids, with protein at 10% the only other major component. Whey may be available for such use either as untreated liquid cheese whey or may be concentrated to a molasses at 70% solids or dry whey powder.

Table 3. The chemical composition of whey (g/l) (Moulin and Galzy, 1984).

Source/coagulant	DM	Lactose	Lipids	Ash	Lactic	N
Cow/rennet	71	52	5	5	0.3	1.5
Cow/mixed	70	51	4	6	2.2	1.5
Cow/lactic ferm.	66	45	1	7	7.5	1.2
Ewe/rennet	84	52	7	6	1.7	2.9
Goat/lactic	63	40	0.4	8	8.7	1.5

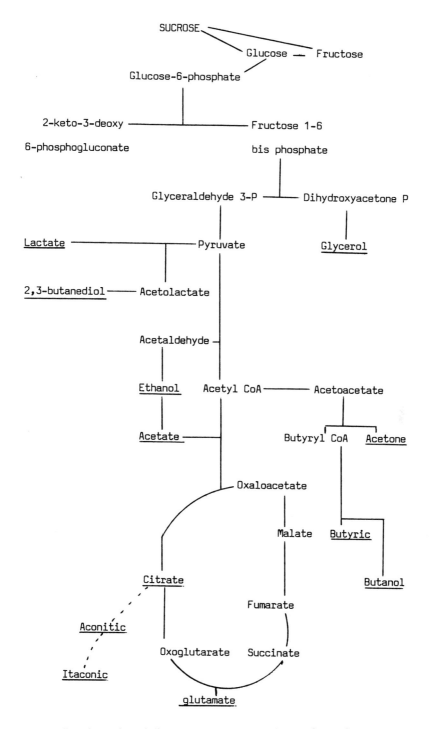

Figure 1. An outline of central metabolism and routes to common fermentation products.

Physiological and biochemical aspects

Metabolic pathways

Almost all commercial bulk fermentation products derived from molasses or whey are by-products of central metabolism through glycolysis, the tricarboxylic acid cycle and related anaerobic or aerobic pathways (*Figure 1*); or are the cell mass itself in the case of baker's yeast and single cell protein.

Glycolysis is the process of metabolism of glucose to lactic or pyruvic acid with the production of ATP. In aerobic conditions the sequence from glucose to pyruvate forms the first part of the respiratory process. In anaerobic conditions pyruvate is reduced to lactate or through acetaldehyde to ethyl alcohol (ethanol). Some bacteria may also convert pyruvate to diacetyl or 2,3-butanediol via acetolactic acid and acetoin. In aerobic respiration the pyruvate is converted to acetyl-CoA which enters the tricarboxylic acid cycle through which citric and glutamic acid may be accumulated to abnormal levels by mutant organisms.

In the absence of oxygen, the disposal of electrons and regeneration of the oxidized electron carriers is a major problem for the cells. In some organisms electrons may be combined with hydrogen in the presence of hydrogenases. However, not all organisms follow this route; in many bacteria organic compounds act as the electron acceptors. The major processes are the reduction of pyruvate to lactate, the reductive decarboxylation of pyruvate to ethanol and the reduction of acetoacetate to butyrate. If the acceptor consumes more electrons than are produced in the form of reduced co-factors, as in butanol formation, acetone is produced in the reaction which produces the extra electrons. Thus acetone production and butanol production are generally inseparable.

Such reactions form the basis of the alcoholic butyric and lactic acid fermentations, including the acetone−butanol fermentation. An exception to this is the production of acetic acid from ethanol which is an aerobic process.

Production of different compounds is also regulated by the presence of key enzymes. Selective inhibition of the enzyme for ethanol formation causes yeast to reduce dihydroxyacetone phosphate to glycerol phosphate, forming glycerol as a final product. Under conditions of low oxygen, substrate may be utilized incompletely. Some electrons are used to reduce oxygen to water and some reduce an organic intermediate to a useful product. An example of this is the reductive decarboxylation of acetolactate to 2,3-butane diol.

Oxygen is the most favourable electron acceptor, and usually under aerobic conditions glucose is oxidized completely to carbon dioxide and water. The route of carbon metabolism, through the tricarboxylic acid cycle, may be inhibited either in mutants which lack key enzymes, or by adjustment of the growth conditions, as in the formation of citric acid, where the enzyme for the further metabolism of citrate is inactivated at low pH.

Utilization of molasses and whey

A wide range of bulk products may be produced by fermentation of molasses (Reed, 1982); the major products are listed in *Table 4*.

Table 4. Fermentation products from molasses and sugar juices.

End use	Product	Substrate	Organism
Beverages	Rum	Molasses/juice	Spontaneous yeast *Saccharomyces*
Food	Bakers' yeast	Molasses	*S. cerevisiae*
	Vinegar	Molasses	*Acetobacter*
Feed	SCP	Molasses/pulping liquor	*S. uvarum, Candida utilis Paecilomyces varioti*
Organic acids	Glutamate	Molasses	*Corynebacterium glutamicum*
	Citric acid	Molasses	*Aspergillus niger*
	Itaconic acid	Molasses	
	Lactic acid	Molasses	*Lactobacillus*
	Acetic acid	Molasses	*Acetobacter*
Enzymes	Glucose isomerase	Xylose	*Streptomyces phaeochromogenes*
	Glucose oxidase	Molasses	*Pencillium vitale*
Gums	Xanthan	Sucrose	*Xanthomonas*
	Dextran	Molasses	*Leuconostoc*
Solvents	Acetone/butanol	Molasses	*Clostridium acetobutylicum*
	Ethanol	All	*S. cerevisiae*
Fuel/octane enhancer	Ethanol	Beet juice Cane juice/molasses	*S. cerevisiae*
	Methane (biogas)	Molasses/stillage	Mixed populations

Products from molasses

(i) *Alcohol.* The production of alcohol (ethanol) grown on sugar by fermentation is mainly carried out using yeasts (*Saccharomyces cerevisiae*) containing substrates which may be directly expressed juice, molasses of any type or whey as well as glucose syrups derived from starch. Products include alcoholic beverages (rum) or the production of alcohol for use as a chemical feedstock, fuel additive or fuel in its own right. The largest use of such fermentation forms the basis of the Brazilian fuel alcohol programme which produces over 11 billion litres per year from sugar cane which is mainly used as a road transport fuel but also as industrial solvent, as precursors for the synthesis of various acids, esters and aldehydes as well as for the production of polymers including acetic acid, ethyl acetate, ethyl ether and polyethylene. Alcohol may also be used in internal combustion engines as fuels, octane enhancers or co-solvents in petroleum blends.

(ii) *Bakers' yeast.* These yeasts are strains of *Saccharomyces cerevisiae* propagated by pure culture methods using cane or beet molasses as substrate under highly aerobic conditions. The yeasts are harvested by centrifugation to produce a cream of about 20% solids, which is pressed in a filter press, or filtered in a rotary vacuum filter to produce bakers' compressed yeast. Alternatively the yeast may be dried.

(iii) *Citric acid.* This is produced commercially by fermentation using fungi such as *Aspergillus niger* grown in aerated submerged culture. Mutant strains are used or metabolic inhibitors added to block the tricarboxylic acid cycle. The world manufacturing capacity is around 200 000 tonnes per annum.

(iv) *Glutamic acid*. This amino acid is produced by fermentation of carbohydrate, often molasses, using any of a number of different bacteria, which although put in separate genera are similar. The main organism used is *Corynebacterium glutamicum* (synonym *Micrococcus glutamicus*). These bacteria require biotin and lack the enzyme oxoglutarate dehydrogenase. At suboptimal levels of growth over 50% of the carbon supplied may be converted directly to glutamate, with little production of any other by-product.

(v) *Acetone−butanol*. Once one of the largest industrial fermentations, based on molasses, using the bacterium *Clostridium acetobutylicum* to produce a mixture of acetone and butanol; now few if any plants remain in production since the advent of lower cost products produced by the petrochemical industry. However, research interest in this process is now increasing as a possible means of producing co-solvents for addition to lead-free petroleum blends and a pilot plant facility to demonstrate such products using new technology is being built in France.

(vi) *Single cell protein*. Large amounts of feed yeast are produced as a by-product of other processes such as brewing and ethanol production. Inactive dried bakers' yeast may be used as a feed supplement. Purpose grown yeasts include *S. cerevisiae*, *S. uvarum* and *C. utilis* produced on molasses. Very large quantities of these are produced in East Europe and Russia.

(vii) *Products from sulphite liquor*. Sulphite liquor from hardwood pulping containing about 3% fermentable sugar has been used as a substrate for the production of SCP using *C. utilis* (Reed, 1973) or the fungus *Paecilomyces varioti* (Romantschuk and Lehtomaki, 1978).

Products from whey

Unhydrolysed whey is a suitable substrate for only a limited number of organisms (see *Table 5*) which are capable of hydrolysing lactose and then using the released sugars in generation of food yeast, SCP or ethyl alcohol as well as partial fermentations which leave a galactose residue. Otherwise the basic fermentations are similar to those which use molasses as a substrate. The first SCP processes were two-stage using lactic bacteria to transform the lactose which was then utilized by *Candida krusei* under aerobic

Table 5. Whey fermentations.

Product	Organism
Food yeast	*Candida utilis/krusei, Kluyveromyces fragilis, Candida intermedia*
Feed supplement	*Lactobacillus bulgaricus, Trichosporon cutaneum, Penicillium cyclopium*
Ethanol	*K. fragilis, C. pseudotropicalis, Saccharomyces* (hydrolysed)
Galactose	*Saccharomyces rosei*
Hydrolysate	*A. niger, K. fragilis* (galactosidase)
Cellulase	*Trichoderma reesei*
Methane (biogas)	Mixed populations

conditions; direct transformation can be achieved using *Kluyveromyces fragilis* or *Candida intermedia*. *K. fragilis* or *C. pseudotropicalis* are also used in the production of ethanol from whey in USA and Ireland.

Anaerobic digestion

Anerobic digestion, with the production of methane, can be a method of pollution control and cut process energy costs through utilization of the gas produced. The process is used to treat a wide range of commercial effluents from food and beverage plants which use molasses as substrate for production of rum, fuel alcohol, citric acid or glutamic acid. Older systems use simple tanks with fairly long residence times. However, worldwide a number of different designs (anaerobic filters, sludge blankets or fluidized bed systems) have been adopted with the objective of maintaining a high active biomass in the reactor whilst passing the aqueous stream through at a high rate. The first such digester in the UK deals with all the whey from the cheese vats (BABA, 1984), as well as effluent from the creamery.

Whey from the cheese vats is piped into bulk storage tanks via a smaller reception tank to the first stage of the modified upflow anaerobic sludge blanket reactor which has a volume of 2000 m^3. An external re-circulation system incorporating heat exchangers holds the temperature at 35°C. The average retention time is about 20 days. During this period, the chemical oxygen demand of the effluent is reduced by between 90 and 95%. Effluent leaving the digester is then treated aerobically to reduce biochemical oxygen demand levels to less than 20 mg/l.

As a result costs of disposal which were previously in the region of £30 000 per annum, have been converted to assets of £80 000 the value of replacement energy. On full output, the biogas generated can have a value of £109 000 per annum (at present prices) as a replacement for heavy fuel oil, and about £60 000 in electrical generation from around 2.5 thousand tonnes of cheese generating up to 100 m^3 of whey per day.

Availablity and costs

The European and UK current and future need for all carbohydrate fermentation feedstocks is estimated in tens of thousands of tonnes if the production of alcohol and

Table 6. Carbohydrate crop production in the EEC (million tonnes) (FAO 1985) 1984 values.

	Million tonnes			
	Wheat	*Barley*	*Beet*	*Potatoes*
Benelux	1.3	0.9	5.7	1.7
Denmark	2.4	6.1	3.2	1.1
France	32.9	11.5	27.8	6.2
Germany	10.2	10.3	20.0	7.8
Greece	2.6	0.8	1.7	1.0
Ireland	0.7	1.6	1.7	1.0
Italy	10.0	1.6	11.2	2.6
Netherlands	1.1	0.2	7.0	6.7
Portugal	0.5	0.1	0.1	1.1
Spain	6.0	10.7	9.1	5.9
UK	14.9	10.9	8.6	7.4

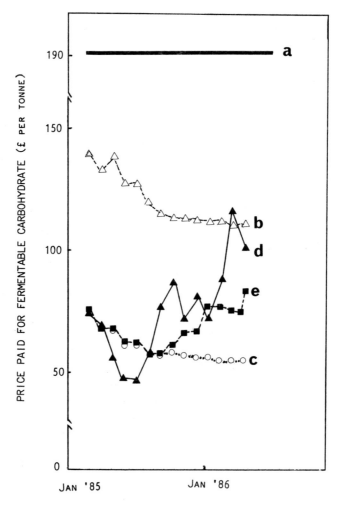

Figure 2. Comparison of the fermentable sugars from various sources during 1985/6. Note, these costs do not include any subsequent processing but merely represent the price paid (£ per tonne, converted where necessary from official figures in $US per pound) for sugar when the indicated commodity was bought. (a) Sugar beet purchased at UK average price to farmers; (b) cane purchased at $US 20 per tonne; (c) cane purchased at $US 10 per tonne; (d) International Sugar Agreement, World Price of Sugar; (e) US molasses price. [Sources of primary data: CSO (1985); Financial Times, London (appropriate dates); FO Licht, International Molasses Report (appropriate dates).]

food products is discounted (CEFIC, 1985). These needs are insignificant as compared with production levels of sugar and starch crops as shown in *Table 6*. These represent a position of 125% self sufficiency for grains and over 140% self sufficiency for sugar, resulting in a carbohydrate surplus in the EEC that runs into millions of tonnes. Hence any question of availability is not one of material supply, but rather of cost.

Whilst the price paid to farmers for beet and the price of sugar are maintained high under the Common Agricultural Policy the overproduction in Europe, as well as in many other countries, has had a serious depressing effect on world prices.

The world sugar market has always had a number of unusual characteristics resulting from the complex nature of the pricing structure which does not necessarily reflect production costs. Prior to 1973 the market was fairly static. However, in 1974 very rapid price rises were followed by a general slump in price and the start of a build up of surplus stocks with prices remaining around $200 per tonne through to 1979. Prices again peaked in 1980 briefly approaching $900 per tonne, but since then have declined dramatically. Although very recently, (*Figure 2*), there has been some price recovery, for much of the last 2 years world prices for raw sugar have remained below production costs, even for the lowest cost cane producer and has been catastrophic for those who do not have a supported or preferential market to sell into. The depressed world market reflects continuing increased production both within the EEC and elsewhere as well as declining consumption in developed countries and an increased share of the sweetener market being taken by high fructose syrup and new high intensity sweeteners such as aspartame.

Over 85% of the world's sugar is traded through preferential agreements resulting in a reasonable return on investment for those associated with them. For instance last year sugar beet prices averaged around £30 per tonne. However, the residual sugar is available on the world market at a variable price, which has been for the last few years very low, generally below production cost. Obviously if sugar can be used for fermentation at these prices and the products freely traded into the EEC, or tariffs are insufficient to negate the advantage of a lower raw material cost then the EEC fermentation industry is obviously at a disadvantage.

Although overproduction of carbohydrate has not proved to be of advantage to the European fermentation industry, with some producers of bulk chemicals moving capacity outside the EEC, there is now pressure on the fermentation industry to act as a means of reducing these surpluses through the production of fuel alcohol (Clarisse, 1986; Coombs, 1985). However, once again the high costs of starch and sugar make this economically unattractive unless massive subsidies occur. Even so it has been suggested by some analysts that such use may be less costly than the alternative of subsidised export from the Community. The eventual answer is not technical, but political. Already, some changes have been made in the sugar and starch regulations in order to lower the cost of fermentation substrate (OJ, 1986; see Chapter 1) for chemical production. It remains to be seen whether such regulations are extended to cover fuel ethanol and if so what this will cost the Community.

References

Alexander,A.G. (1985) *The Energy Cane Alternative*. Elsevier, Amsterdam.

BABA (1984) Anaerobic digesters treating industrial effluent in the UK. *The Digest,* **15**, 6−9.

Baker,B.P. (1982) Cane and beet molasses and related products — quality control. In *Molasses the Management of Quality*. United Molasses Company, London.

Campbell,G.K.G. (1976) Sugar beet. In *Evolution of Crop Plants*. Simmonds,N.W. (ed.), Longman, London and New York, pp. 25−28.

Clarisse,Y. (1986) *The Future of Bioethanol in Europe*. European News Agency, Bruxelles.

CEFIC (1985) *Use of Agricultural Raw Materials in the European Chemical Industry*. CEFIC, Bruxelles.

Coombs,J. (1985) Outlook for oxygenates in the European market, the role of ethanol. In *International Symposium on Sugar and Alcohol*. Copersucar, Sao Paulo, pp. 241−256.

CSO (1985) *Annual Abstracts of Statistics*. Central Statistical Office, HMSO, London.

FAO (1985) *Year Book of Statistics*. Food and Agricultural Organisation, Rome.

Lome Convention (1986) In *The European Community*. (2nd Edition) Morris,B. and Boehm,K. (eds.), Macmillan Press, London p. 302.

Moulin,G. and Galzy,P. (1984) Whey a potential substrate for biotechnology. *Biotechnol. Genet. Eng. Rev.*, **1**, 347−374.

OJ (1986) *Official Journal of the European Communities*. L94, Council Regulations 1006/86−1009/86. Commission of the European Communities. Bruxelles.

Parker,K.J. (1982) Background and sources of molasses. In *Molasses the Management of Quality*. United Molasses Company, London.

Paturau,J.M. (1982) *Byproducts of the Cane Sugar Industry*. 2nd Edition, Elsevier, Amsterdam.

Reed,G. (1973) *Yeast Technology*. AVI Publishing Company Inc., Connecticut.

Reed,G., ed. (1982) *Prescott and Dunn's Industrial Microbiology*. 4th Edition, AVI Publishing Company Inc., Connecticut.

Romantschuk,H. and Lehtomaki,M. (1978) Operational experiences of first full scale Pekilo process SCP-mill application. *Process Biochem.*, **13**, 16−17.

CHAPTER 4

Utilization of cellulose as a fermentation substrate: problems and potential

J.WOODWARD

Chemical Technology Division, Oak Ridge National Laboratory, Oak Ridge, TN 37831, USA

Introduction

Since 1973, when the vulnerability of industrialized and developing nations to limited supplies of crude oil became apparent, much effort has gone into finding alternatives to oil as a source of fuels and chemicals. One such alternative, cellulose, is not only the most abundant form of carbon in the world, but is a rapidly renewable resource – unlike fossil fuels such as crude oil. Typical sources of cellulose include hardwood and softwood trees, agricultural crop residues and municipal wastes. All of these have the potential to be fermented into a variety of the types of fuels and chemicals which are currently obtained from oil.

Cellulose can be utilized directly or indirectly as a fermentation substrate. Direct cellulose utilization is achieved by microbial growth on a cellulosic substrate, and involves a one-step conversion of cellulose to ethanol by mixed cultures of *Clostridia* sp. Indirect cellulose utilization involves the use of enzymes or acids that hydrolyse cellulose to glucose. The resulting hexose can be subsequently fermented to ethanol (Bungay, 1984). The problems and potential in using either method for glucose utilization will be discussed.

Widespread acceptance of cellulose as a fermentation substrate has not yet been commercially realized, primarily because it is presently cheaper to produce ethanol from oil than from cellulose. Furthermore, there now exists a surplus of oil in the world market that has resulted in a dramatic decrease in its price. At the present time, therefore, there appears little interest in, or incentive for, developing the technology for the fermentation of cellulose on an industrial scale. Needless to say, however, this picture could change dramatically in the event of circumstances leading to further oil shortages and price rises. We must be prepared for such an event and continue the research and development that will result in economic processes for the utilization of cellulose as a fermentation substrate. For example, Mandels (1985) has outlined existing research opportunities which could improve the economics of enzymatically hydrolysing cellulose to glucose. These opportunities include: (i) development of new strains of microbes that hyperproduce cellulase; (ii) improvement of methods for the pre-treatment of cellulosic materials that is necessary for complete utilization; (iii) work on the stabilization, recovery, and re-use of cellulase enzymes; and (iv) research to complete our understanding of how cellulase acts upon cellulose and, in particular, to elucidate the mechanism of synergism amongst the individual cellulase components.

Some of the current research in these and other areas will be discussed, with particular emphasis on the problems to be addressed and the potential for solving them.

Pre-treatment of cellulosic substrates

Cellulose from sources such as hardwood and softwood trees, and agricultural wastes (for example, wheat straw, bagasse and corn residues), exists in a lignin−cellulose−hemicellulose complex. The efficiency of enzymatic hydrolysis of cellulose in such a complex to form glucose is low unless the complex is broken up by its pre-treatment prior to hydrolysis. Pre-treatment of such lignocellulosic materials renders them more susceptible to enzymatic hydrolysis by increasing the number of glycosidic bonds accessible to the enzyme. Such pre-treatment is also essential for the direct utilization of cellulosic materials by anaerobic bacteria. It is not, however, as important in the acid hydrolysis of lignocellulose, because the acid can easily permeate the complex (Nystrom *et al.*, 1984).

An indication of the efficiency of these pre-treatment methods may be obtained by comparing the percentage of saccharification of cellulose to glucose in samples before and after pre-treatment. For example, Vallander and Eriksson (1985) compared the degree of saccharification of wheat straw achieved by pre-treatment with steam explosion, defibration and the use of alkaline H_2O_2 with that obtained without pre-treatment (*Table 1*). Saccharification was achieved using a commercial preparation of cellulase from *Trichoderma reesei*. Steam explosion is seen to be the most effective method of pre-treatment in this case. In this regard, the Iotech process of steam exploding wood chips can be considered an applicable technology. In this process, cellulosic material is heated under high-pressure steam and exploded by the rapid release of pressure. This causes melting of the lignin and reduces the degree of polymerization of cellulose, which is subsequently easily hydrolysed by cellulase.

In other recent work, Gould and his colleagues at the US Department of Agriculture in Peoria, Illinois, have studied the effect of dilute alkaline solutions of H_2O_2 on the susceptibilty of agricultural residues to enzymatic and microbial degradation (Gould,

Table 1. Degree of saccharification of wheat straw after different pretreatments.

Pretreatment	Degree of saccharification (%)[a]				Total sugar yield[b] (mg/g)
	Pretreatment		Enzymic hydrolysis		
	Reducing sugar	Total sugar	Reducing sugar	Total sugar	
Steam explosion	5 −13	14−28	18−75	41−92	290−650
H_2O_2/OH⁻	1 −2	6−23	18−56	22−71	155−500
Defibration	0.5−2	4−8	16−29	26−50	185−355
Control	−	−	10	12	85

[a]Based on theoretical amounts to be recovered from wheat straw.
[b]Total sugar yield in mg/g of wheat straw.
Reproduced from Vallender and Eriksson (1985), with permission. Copyright John Wiley and Sons, Inc.

Table 2. Effect of alkaline hydrogen peroxide treatment on the saccharification efficiency of various lignocellulosic materials.

Material	Saccharification efficiency (%)[a]	
	$- H_2O_2$	$+ H_2O_2$
Wheat straw	27	93
Corn stalks	49	100
Corn cobs	32	100
Corn husks	62	99
Foxtail	27	82
Switchgrass	52	81
Blue bluestem	46	97
Cattail	34	60
Alfalfa	41	94
Soybean stover	46	75
Kenaf	26	58
Oak	22	53
Sugar cane bagasse	28	95
Peanut hulls	12	40

[a]Samples were treated at pH 11.5 ± 1% hydrogen peroxide for 24 h at 25°C, dried and treated with cellulase. Data presented at the 190th Meeting of the American Chemical Society, Chicago, Illinois, September 8–13, 1985, by Gould *et al.*, 1985.

1984; Gould *et al.*, 1985). Their data (in *Table 2*) show that the saccharification efficiency of cellulase in various agricultural residues was markedly increased by pre-treatment of the cellulosic materials with 1% (w/v) H_2O_2 at pH 11.5 for 24 h at 25°C. During the pre-treatment of wheat straw, approximately 50% of the lignin content was solubilized, and the presence of hemicellulose did not affect the efficiency of cellulose saccharification. A major problem with this method of pre-treatment is the cost of the H_2O_2, which is US$0.22 per 0.33 kg of lignocellulose. However, potential exists to reduce the cost, if divalent metals and catalase are removed from the substrate prior to pre-treatment, thereby eliminating wasteful decomposition of the peroxide and allowing a substantial reduction in its requirement.

Other methods for the pre-treatment of lignocellulosics recently described include ammonia-freeze explosion (Dale *et al.*, 1985) and the use of a cellulose solvent composed of a ferric sodium tartrate complex in 1.5 M NaOH (Hamilton *et al.*, 1984). Both of these methods significantly enhanced the saccharification of a variety of lignocellulosics. Dale and his colleagues calculated the operating costs for the ammonia-freeze explosion process to be (US) approximately $20 per tonne of wheat straw (US $0.02 per kg, where 1 tonne = 1000 kg). Hamilton *et al.* claimed that almost complete recovery of the solvent is possible by washing the treated corn residue with 20 volumes of water per volume of solvent.

An interesting example of simultaneous pre-treatment and saccharification is provided by the work of Lee and his co-workers at Washington State University (Ryu and Lee, 1983; Deeble and Lee, 1985). In this work, an attrition bioreactor was used in which stainless steel balls were mixed with the cellulosic substrate during enzymatic hydrolysis. As shown in *Figure 1*, the percentage conversion of newsprint to a reducing sugar was higher in the attrition bioreactor than when the reaction was carried out in the absence of the stainless steel balls. It appeared that the simultaneous milling and hydrolysis was

47

J. Woodward

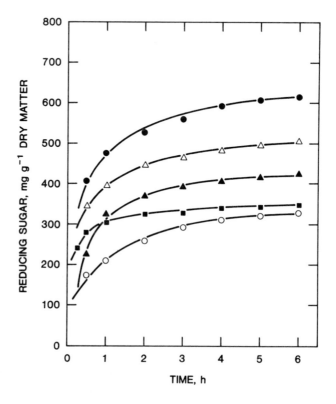

TIME, h

Figure 1. Hydrolysis of newsprint; in a regular stirred reactor without pre-treatment (○), and after 1 week of dry ball milling of the substrate (■); hydrolysis in an attrition bioreactor with sand (▲), with 0.318 cm diameter stainless-steel balls (△) and with 0.476 cm diameter stainless-steel balls (●), all with an agitation speed of 400 r.p.m. Reproduced from Ryu and Lee (1983) with permission. Copyright John Wiley and Sons, Inc.

more effective than the combination of separate milling and hydrolysis operations. Although cellulase activity was 55% inactivated after 2 h in the attrition bioreactor, complete stabilization of the enzyme was achieved when the surfactant Triton X-100 was present in the reactor at a concentration of 1 mg/ml. The stabilizing effects of surfactants on cellulase activity are described later in this chapter.

There are two other intriguing possibilities for the delignification of lignocellulosic materials that can be mentioned. First, the separation and characterization of ligninases from the culture fluid of the wood-decomposing basidiomycete *Phanerochaete chrysosporium* have been described (Tien and Kirk, 1983; Kuwahara *et al.*, 1984; Kirk *et al.*, 1986). These oxidative enzymes are H_2O_2-requiring haemoproteins that degrade lignin and lignin model compounds and thus could prove to be useful for simultaneous delignification and saccharification of lignocellulosic substrates. Compatibility of the properties of cellulase and ligninase would be important for such a process. Second, inhibitors of lignin biosynthesis (Amrhein *et al.*, 1977, 1983; Hawkins *et al.*, 1981) could be used for the production of lignin-free grasses. One such inhibitor, amino-oxy-β-phenylpropionic acid, inhibits a key enzyme in the phenylalanine−cinnamic acid

48

Table 3. Advantages and disadvantages of using acid for the saccharification of lignocellulosic materials.

Advantages

1. Pre-treatment of feedstock not necessary.
2. Short reaction times.
3. Cheap and readily available catalyst.
4. Low temperatures with concentrated acid.

Disadvantages

1. Concentrated acid recovery essential but expensive.
2. High temperatures (180°C) required with dilute acid.
3. Sugar product decomposition to furfural and organic acids.
4. Neutralization required before fermentation of hydrolysates.
5. By-products of hydrolysis toxic to yeast.
6. Lower potential yield of ethanol.
7. Expensive construction materials necessary.

pathway for the biosynthesis of lignin. Current research at ICI's plant production division at Jeallot's Hill in Berkshire, UK, is aimed at the production of lignin-free grasses. Clearly, such grasses would have improved susceptibility to digestion by cellulase.

Finally, it should be mentioned that the mechanism whereby efficient utilization of lignocellulose occurs after pre-treatment appears to be due to an increase in pore size of the lignocellulosic substrate and is not related to crystallinity of the substrate (Puri, 1984; Grethlein, 1985). Klyosov (1985) also has shown that crystallinity does not affect the extent to which cellulase adsorbs to cellulose.

Acid hydrolysis of cellulose

The advantages and disadvantages of using acid to hydrolyse cellulose are shown in *Table 3*. Ladisch and Tsao (1986) have, however, described the development of a low-temperature process for cellulose hydrolysis in which the lignocellulosic substrate is pre-treated with concentrated sulphuric acid prior to hydrolysis using dilute sulphuric acid at 100°C. Glucose yields of 80−95% (w/v) are obtained with little degradation. Also, the acid is recycled in this process. They have estimated that the cost of cellulose conversion to ethanol using acid hydrolysis is approximately US$1.10 per gallon.

It is of interest, when considering acid hydrolysis of cellulose, to mention the work at ICI's agricultural division in which metal salts are used in combination with acid at 50−90°C for the rapid (5−30 min) hydrolysis of cellulose to glucose (80% w/v yields) (A.J.Beardsmore, personal communication).

Enzymatic hydrolysis of cellulose

The advantages and disadvantages of using enzymatic hydrolysis of cellulose to provide fermentable sugar are shown in *Table 4*. Much has been written concerning the enzymatic hydrolysis of cellulose by cellulase enzymes and many colloquia have been organized that deal solely with this subject. Wood (1985) has reviewed the properties of cellulases from fungi such as *T. viride, T. reesei, T. koningii, Penicillium funiculosum* and *Talaromyces emersonni.*

49

Table 4. Advantages and disadvantages of using enzymes for the saccharification of lignocelluosic materials.

Advantages

1. Mild conditions of temperature and pH.
2. No degradation products.
3. Simultaneous saccharification/fermentation possible.
4. Expensive corrosion-resistant reactors not required.

Disadvantages

1. Pre-treatment of lignocellulosic substrate essential.
2. Products of saccharification inhibit hydrolysis.
3. High levels of enzyme required.
4. Adsorption and loss of enzyme on undigested residues.
5. No satisfactory method for recovery and recycle of enzyme.

Cellulases from these fungi are extracellular proteins containing three major active components referred to as:

(i)   endo-1,4-β-glucanase (EC 3.2.1.4);
(ii)   cellobiohydrolase (EC 3.2.1.91);
(iii) β-glucosidase (EC 3.2.1.21).

It is generally accepted that the endoglucanase and cellobiohydrolase adsorb onto the surface of insoluble crystalline cellulosic substrates, resulting in depolymerization of the chain length and splitting off of cellobiose units from the non-reducing end of the cellulose molecule, respectively. The actions of these two components result in the formation of cellobiose, primarily, as well as some glucose. The β-glucosidase component completes the hydrolysis of cellobiose to glucose.

Considering the enzymatic hydrolysis of cellulose under use conditions, Mandels *et al.* (1981) have shown that whereas dilute cellulase activity [0.625 filter paper units (FPU)] yielded about 12% conversion of 15% (w/v) avicel to reducing sugars in 24 h, 160 FPU were required to achieve about 30% conversion. This means that a 256-fold increase in enzyme concentration is needed to increase the percentage of conversion by only a factor of 2.5. They found this to be true even for ball-milled newspaper (*Figure 2*). Therefore, the indirect utilization of cellulose as a fermentation substrate requires large quantities of enzyme if a large percentage of conversion of cellulose to fermentable sugars is to be achieved.

The two approaches that can be used to solve the problem of the large enzyme requirement are:

(i)   the development of new strains and mutants of microbes with higher specific cellulase activities than those currently obtained; and
(ii)   the development of a suitable method to recover and recycle the enzymes so that they can be used repeatedly for cellulose hydrolysis.

Solutions to these problems would be major advances, since enzyme production costs account for as much as 60% of the total processing costs associated with the enzymatic hydrolysis of cellulose to glucose (Deshpande and Eriksson, 1984).

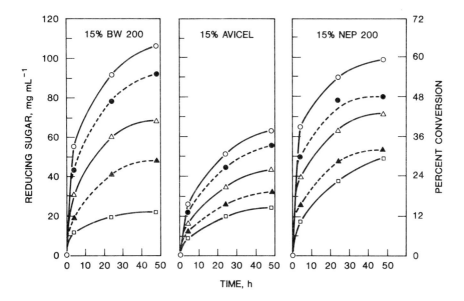

Figure 2. Effects of time and cellulase concentration on the hydrolysis of cellulosic substances. 15% BW 200 with ball-milled sulphite pulp; avicel, with microcrystalline cellulose; and 15% NEP 200 with ball-milled newspaper. Conditions: pH 4.8, 50°C, (○——○) 160 units; (●- - -●) 40 units; (△——△) 10 units; (▲——▲) 2.5 units; (□——□) 0.75 units. Reproduced from Mandels *et al.* (1981), with permission. Copyright John Wiley and Sons, Inc.

Most of the work on the selection of hypercellulolytic mutants has been done starting with the wild-type *T. reesei* QM6a. Mutagenesis has led to the development of strains that do produce higher titres of cellulase activity (as shown in *Table 5*). Especially noticeable are the two strains Rut-NG14 and L27, produced in the laboratories of Montenecourt and Shoemaker, respectively (Montenecourt, 1983; Shoemaker *et al.*, 1981). However, no dramatic increases of specific activity towards filter paper, were observed. In fact, a decrease in the specific activity of endoglucanase is seen in many mutants. Perhaps, as the development of new cellulolytic strains of fungi or bacteria continues, higher specific activity towards insoluble lignocellulosic substrates will be obtained. Another possibility is the chemical modification of existing cellulases to artificially increase their specific activities. However, an effort towards chemical modification of the free ϵ-amino groups of *T. reesei* C30 cellulase with monofunctional and bifunctional imidoesters resulted in decreased activity and decreased stability of the enzyme (Woodward *et al.*, 1981). It should be mentioned that site-directed mutagenesis may, in the future, prove very useful in improving the activity and stability of cellulase.

Recovery and re-use of cellulase

Ideally, an immobilized cellulase could be used to hydrolyse lignocellulosic substrates, recovered from the reaction medium and used repeatedly. This would make the economics of cellulose utilization using cellulase much more favourable. However, the

Table 5. Yield of cellulase and specific activity of various *T. reesei* mutants.

Strain	Substrate	FPU (IU/ml)	Endoglucanase (IU/ml)	Soluble protein (mg/ml)	Specific FPU activity
QM6a	6% compressed milled cotton	5.0	88	7.4	0.67
QM9414	6% compressed milled cotton	10.0	109	13.6	0.73
MCG77	6% compressed milled cotton	10.7	104	16.2	0.66
MCG80	9% avicel pH 102	15	−	22.7	0.66
Rut-NG14	6% compressed milled cotton	14.8	133	21.2	0.70
Rut-C30	6% compressed milled cotton	13.6	181	20.6	0.66
Rut-C30	5% bleached pulp (BW200) cornsteep liquor	12.2	225	8.5	1.43
RL-P37	5% bleached pulp (BW200) cornsteep liquor	10.1	356	7.6	1.33
L-27	8% avicel	18	79	22	0.8
VTT-D-79124	3% finnfloc birch cellulose	4.4	78[a]	8.4	0.52
CL-847	5% Whatman CC41 2% wheat bran	14.6	−	21	0.70

[a]Assayed employing hydroxyethylcellulose. All other strains were assayed employing carboxymethyl cellulose. Reproduced from Montenecourt (1983), with permission.

use of an immobilized enzyme to catalyze the hydrolysis of an insoluble substrate does not appear to be logical, because effective interaction between enzyme and substrate would be greatly impaired by the enzyme's immobility.

There are, however, some reports describing the use of immobilized cellulase to hydrolyse insoluble cellulose. For example, Karube *et al.* (1977) immobilized cellulase in a collagen fibril matrix. Although details on the recovery of activity after immobilization were not reported, the immobilized enzyme (in a fluidized bed reactor) reportedly hydrolysed insoluble microcrystalline cellulose (avicel, 0.33% w/v) circulating through the bed. Greater than 80% hydrolysis of the substrate was achieved after 160 h at 30°C. Furthermore, no leakage of the enzyme was reported to occur, which suggests that intimate association between the immobilized enzyme and insoluble cellulose had to occur somehow, even during constant circulation of substrate. Woodward and Zachry (1982) immobilized the cellulase complex of *T. reesei* C30 by covalent coupling to cyanogen bromide-activated Sepharose after aminoalkylation of the carbohydrate side chains of the enzyme with ethylenediamine. More than 90% of the avicelase activity was lost when the cellulase was immobilized by this method. Similar results have been obtained by Fadda *et al.* (1984). Rogalski *et al.* (1985) attached

Table 6. Immobilized cellulase preparations.

Support	Substrate (% w/v)	Leakage	Recovery of activity (%)	Hydrolysis of substrate (%)	Reference
Collagen	Avicel (0.33)	None	n.d.[a]	>80 (160 h, 30°C)	Karube *et al.* (1977)
CNBr−Sepharose	Avicel (2)	None	8	~6 (28 h, 50°C)	Woodward and Zachry (1982)
CNBr−Sepharose	Alfalfa (3)	n.d.	20	~60 (6 days, optimal temperature)	Fadda *et al.* (1984)
Controlled-pore glass	Avicel (1)	n.d.	72	~25 (72 h, 50°C)	Rogalski *et al.* (1985)

[a]n.d. = not determined.

cellulase from *Aspergillus terreus* to controlled pore glass and claimed 72% retention of activity. Some of the data on immobilized cellulases provided by these authors is summarized in *Table 6*.

When using an immobilized cellulase to hydrolyse insoluble cellulose, it is essential to ensure that leakage of enzyme from the support does not occur, otherwise hydrolysis will obviously be caused by the soluble enzyme. This has been shown to be true for cellulase immobilized by adsorption onto concanavalin A attached to an inorganic kieselguhr-based support (Woodward *et al.*, 1986). Thus, if the immobilized cellulase is mixed with insoluble cellulose followed by removal of the support from the reaction medium, hydrolysis by cellulose continues indicating that the enzyme has desorbed from the support when the latter is contacted with cellulose. This phenomenon led to a novel concept for the recovery of cellulase from reaction mixtures, especially when cellulose is used directly as a fermentation substrate (*Figure 3*). Complete recovery of the cellulase is dependent upon two factors:

(i)   complete hydrolysis of cellulose and
(ii)   re-attachment of the cellulase in the presence of ethanol.

In a practical system, however, it is unlikely that complete hydrolysis of a cellulosic substrate will occur, and greater than 50% of the original enzyme used in a saccharification will remain in the adsorbed state, even when there is a high percentage of conversion of the substrate. Elution of the cellulase from the undigested cellulose is essential if it is to be recovered and re-used. Of the three major active components associated with the cellulase enzyme complex, the β-glucosidase is relatively easy to recover, because this enzyme is not adsorbed by cellulose to any great extent. In fact, β-glucosidase can be used in an immobilized form, since its substrate cellobiose is soluble (Lee and Woodward, 1983). Endoglucanase that is adsorbed to cellulose can be recovered by elution of the undigested substrate with a phosphate buffer of pH 7.0 (Sinitsyn *et al.*, 1983) and dilute alkali (Reese, 1982a). Cellobiohydrolase is strongly adsorbed to cellulose and, although urea (6 M), guanidine HCl (4 M), dimethylsulphoxide (3 M), and *n*-propanol (4 M) are good eluants, their use results in inactivation of enzyme activity (Reese, 1982a).

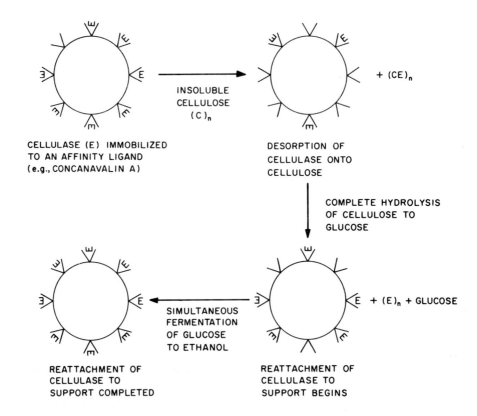

Figure 3. Concept for the recovery of soluble cellulase from reaction mixtures.

Other methods have also been considered for the recycling of cellulase. Ohlson *et al.* (1983) described the enzymatic hydrolysis of cellulose in an ultrafiltration membrane reactor in which the enzymes are retained in the reactor while the soluble sugars permeate the membrane. The main problem of this method is that the enzyme remains adsorbed on undigested substrate. The use of a two-phase, aqueous system for cellulase recovery has also been considered (Tjerneld *et al.*, 1985). In a two-phase system based on the use of dextran, polyethyleneglycol and buffer, the cellulase and cellulose were primarily confined to the lower phase, and the soluble sugars have greater affinity for the upper phase. As with the ultrafiltration method, continuous separation of product is achieved.

Finally, the question of enzyme stability must be considered in relation to recovery and re-use. It is clear that no useful purpose will be served if inactivated enzyme is recovered. Reese (1982b) has shown that shaking inactivates both cellobiohydrolase (avicelase) and endoglucanase activity, although the latter component is much more stable (*Figure 4*). This inactivation can be prevented by including several nonionic surfactants in the reaction medium. *Figure 5* shows the protective effect of polyethyleneglycol 6000 on avicelase at several pH values. Reese also showed that, at 50°C, the half-life of cellobiohydrolase when shaken is 1.5 h, whereas, in the presence

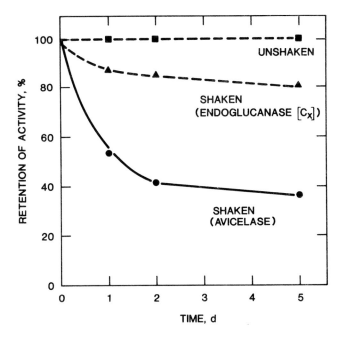

Figure 4. Effect of shaking on inactivation of *T. reesei* cellulase C30 enzymes. Conditions: 30°C, pH 4.9, 300 r.p.m. in a 1 inch circle, 0.2 mg NaN$_3$/ml (preservative). (■): unshaken control samples (both C$_x$ and avicelase); (●): shaken, avicelase; (▲): shaken, C$_x$. Reproduced from Reese (1982b,) with permission.

of the fluorinated surfactant zonyl N, the half-life is increased to 180 h. In a saccharification process at 50°C for 48 − 72 h, the presence of an appropriate surfactant will allow most of the enzyme activity to be recoverable. Interestingly, the protective effect of surfactants is achieved at low concentrations (0.01 − 0.1 mg/ml), which do not inactivate cellulase.

Inhibition of cellulose utilization

Both direct and indirect utilization of cellulose as a fermentation substrate are subject to inhibition. In processes using indirect utilization, whereby cellulose is enzymatically hydrolysed to sugars which are then fermented, the cellulase is inhibited by cellobiose and glucose. More specifically, cellobiose is a potent inhibitor of cellobiohydrolase (Mandels, 1985) and also inhibits the high molecular weight endoglucanase of *T. koningii* (Churilova *et al.*, 1980). This is a major problem for *Trichoderma* cellulases because their levels of β-glucosidase are too low to be of practical use. Consequently, cellobiose accumulates and inhibits the action of cellulase. This problem can be solved easily by supplementing reaction mixtures containing cellulose and *T. reesei* cellulase, for example, with β-glucosidase from a source such as *Aspergillus niger*. It is known that the *Aspergillus* spp. are superior to other microorganisms in their production of β-glucosidase that is highly active on cellobiose (Woodward and Wiseman, 1982). Furthermore, since cellobiose is soluble, β-glucosidase can be added in an immobilized

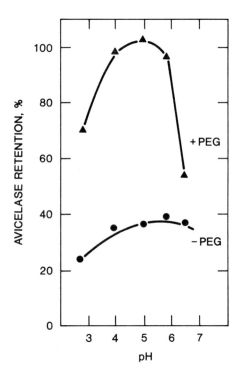

Figure 5. Effect of pH on the protective effect of PEG 6000 against the inactivation of avicelase by shaking. Conditions: as in *Figure 4*; time, 5 days, PEG at 0.2 mg/ml. Reproduced from Reese (1982b), with permission.

form that can be easily recovered and re-used. The data shown in *Figure 6* are an example of the effect of adding β-glucosidase (immobilized on alumina) to a mixture of cellulose (solka floc) and cellulase. It is clear that, in the presence of immobilized β-glucosidase, both the yield and rate of glucose production are enhanced. Also, since the concentration of cellobiose is lower, especially after prolonged hydrolysis times, inhibition of cellobiohydrolase will be minimal (Sundstrom *et al.*, 1981).

Besides the product inhibition of cellobiohydrolase and endoglucanase by cellobiose, substrate inhibition of *A. niger* β-glucosidase by cellobiose occurs at concentrations >10 mM (Lee and Woodward, 1983; Grous *et al.*, 1985). At lower concentrations, no problem exists. Lee and Woodward (1983) showed, however, that when β-glucosidase from *A. niger* immobilized on concanavalin A – Sepharose (CAS) was entrapped within calcium alginate gel spheres, the gel-entrapped enzyme was not subject to substrate inhibition by up to 100 mM cellobiose, unlike the immobilized enzyme (*Table 7*). Clearly, the restriction of substrate diffusion is the explanation of this result, which could have important practical consequences for the hydrolysis of concentrated cellulose-derived cellobiose solutions.

The β-glucosidase component of cellulase is inhibited by glucose (Woodward and Arnold, 1981); thus, as glucose accumulates during an enzymatic hydrolysis of cellulose, the cellobiose will also accumulate and inhibit the other cellulase components. This

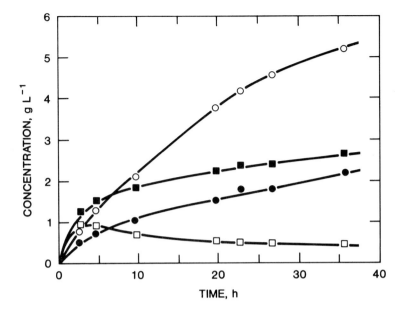

Figure 6. Effect of the presence of immobilized β-glucosidase on the enzymatic hydrolysis of cellulose (solka floc). Conditions: pH 4.8, 50°C, 55 g/l substrate, 220 units *T. reesei* cellulase; 80 units immobilized β-glucosidase (*A. phoenicis*). (○) Glucose, immobilized β-glucosidase added; (■) cellobiose, no β-glucosidase added; (●) glucose, no β-glucosidase added; (□) cellobiose, immobilized β-glucosidase added. Reproduced from Sundstrom *et al.* (1981), with permission. Copyright John Wiley and Sons, Inc.

Table 7. Effect of substrate concentration on effectiveness factor of gel-entrapped immobilized β-D-glucosidase.

Concentration of cellobiose (mM)	Activity of CAS-immobilized enzyme (units)	Activity of entrapped CAS-immobilized enzyme (units)	Effectiveness factor[a] (%)
2	42	10	24
3	53	11	21
5	54	18	33
10	71	25	35
50	42	34	81
100	25	46	184

[a]Activity of entrapped CAS-immobilized enzyme divided by the activity of CAS-immobilized enzyme × 100 = the effectiveness factor.
Reproduced from Lee and Woodward (1983), with permission. Copyright John Wiley and Sons, Inc.

problem can be easily overcome by the removal of glucose as it is formed. This can be done if fermentation of the product glucose to ethanol is carried out simultaneously (Spangler and Emert, 1986). In such a system, glucose does not accumulate but is fermented to ethanol as it is formed. Woodward and Arnold (1981) also suggested that since fructose was a poor inhibitor of β-glucosidase, a mixture of cellulase and glucose isomerase with compatible pH would result in a greater conversion of cellulose to

Table 8. Inhibition of cellulase during a 3-day saccharification.

Additive	Concentration of additive at 50% inhibition (%)
Glucose	2.5
Ethanol	2.6
Acetone	4.4
Butanol	1.7
2,3-Butanediol	4.7
Glycerol	4.6
Acetic acid	>7
Butyric acid	3.1
Citric acid	>7
Sodium gluconate	3.8
Itaconic acid	>7
α-Ketoglutaric acid	>7
Lactic acid	>7
Propionic acid	6.1
Succinic acid	>7

Reproduced from Takagi (1984), with permission. Copyright John Wiley and Sons, Inc.

fermentable sugar (in this case, fructose).

Besides the inhibition of cellulase activities by substrates and by the products of the reaction they catalyse, consideration must also be given to the inhibiting action of fermentation products, especially in simultaneous saccharification/fermentation processes. Takagi (1984) has investigated the inhibition of *T. reesei* cellulase by typical fermentation products. The product concentration required to give 50% inhibition of cellulase (as measured by the production of reducing sugar during a 3-day saccharification) is given in *Table 8*. Butanol, ethanol and glucose are the strongest inhibitors. In related investigations, Sinitsyn *et al.* (1982) found that there were inhibitors of *T. reesei* cellulase in raw steam-exploded wood. The inhibitors could be removed by washing with water, however, and are not a problem. The nature of the inhibitors was not addressed in this research.

Direct utilization of cellulose as fermentation substrate

Direct utilization of cellulose as a fermentation substrate occurs when cellulase production, cellulose hydrolysis and fermentation occur simultaneously in a single operation. One example of such an operation involves the use of the anaerobic, thermophilic bacterium *Clostridium thermocellum* (Cooney *et al.*, 1978). In this study, it was shown that ball-milled corn residue or solka floc could be used as substrates for the growth of this bacterium. The maximum concentrations of the two major products, acetic acid and ethanol, reached during the fermentation were 1.2 and 0.8 g/l, respectively, when the substrate was corn residue. When solka floc was used as the substrate, the concentration of both major products reached 4 g/l. For direct cellulose fermentation to become technically feasible, the amount of accumulated ethanol must be increased. A major problem associated with this process is the sensitivity of cell growth to inhibition by ethanol and acetic acid.

Another disadvantage of direct cellulose fermentation is the generation of unwanted

Table 9. Ethanol and acetic acid production by direct cellulose fermentation.

Microorganism	Substrate (g/l)	Ethanol produced (g/l)	Acetic acid produced (g/l)	Reference
C. thermocellum ATCC 27405	Solka floc (47.5) Corn residue (9.5)	4.0 0.8	4.0 1.2	Cooney et al. (1978)
Clostridium sp. Strain T$_x$	Cellulose pulp (17.5) Rice straw (50.0)	4.0 4.0	1.7 5.3	Taya et al. (1984)
Clostridium sp. Strain H10	Solka floc (6.0)	2.0	8.0	Ford et al. (1984)
Clostridium and Bacillus wild co-culture	microcrystalline cellulose (20.0)	5.1	–	Volfová et al. (1985)

Table 10. Sugar utilization and ethanol and acetic acid production by *Clostridium saccharolyticum* and *Zymomonas anaerobia* alone and in the co-culture[a].

Microorganism	Incubation time (h)	Glucose used (g/l)	Cellobiose used (g/l)	Xylose used (g/l)	Ethanol produced (g/l)	Acetic acid produced (g/l)
C. saccharolyticum	24	7.9	0.1	0.3	2.5	1.2
	48	13.7	0.3	2.9	6.5	1.9
	72	13.8	3.2	6.7	7.6	3.2
	96	13.7	4.1	6.8	7.9	3.6
Z. anaerobia	12	13.4	0	0	5.6	0.1
	24	13.8	0.1	0.1	5.8	0.1
	48	13.9	0.2	0.3	6.0	0.1
C. saccharolyticum plus Z. anaerobia	12	13.7	0.1	0.9	5.8	0.1
	48	13.8	0.3	4.8	7.8	0.2
	72	13.8	4.4	6.3	9.9	1.5
	96	13.7	6.3	6.6	10.9	1.7

[a]The media contained glucose (14 g/l), cellobiose (7 g/l) and xylose (7 g/l).
Reprinted from Asther and Khan (1984), with permission. Copyright John Wiley and Sons, Inc.

by-products. If ethanol is the desired product, its yield per weight of substrate will be less if other products are also generated during the fermentation. The concentrations of ethanol and acetic acid generated by a *Clostridium* sp. fermentation are dependent on the type of substrate utilized. It can be seen from the data in *Table 9* that only low concentrations of these products are produced. In contrast, ethanol concentrations exceeding 100 g/l can be achieved in batch culture when fermentations with a yeast such as *Saccharomyces carlsbergensis* or the bacterium *Zymomonas mobilis* are carried out using glucose as the substrate (Rogers *et al.*, 1980). These microorganisms unfortunately do not utilize cellulose directly.

One approach that can be taken to overcome the inefficient ethanol production of these cellulosic microbes, is the use of mixed cultures. For example, Asther and Khan (1984) used two microbes in culture: *Clostridium saccharolyticum* and *Zymomonas anaerobia*. Although *C. saccharolyticum* produces both ethanol and acetic acid when

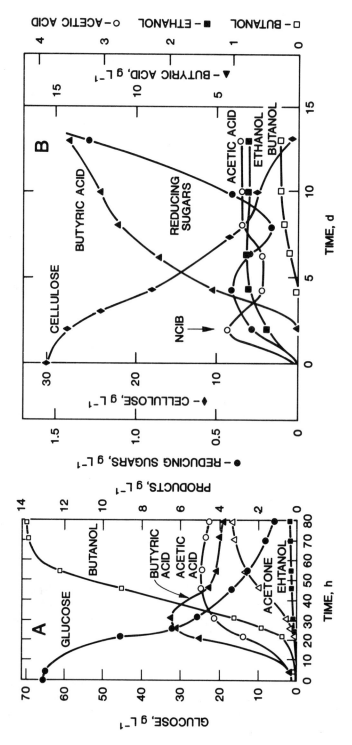

Figure 7. (a) Kinetics of a pH-controlled batch fermentation of *Clostridium acetobutylicum* NCIB 619 on glucose as carbon source. Experiment was performed in a fermentor that contained 1 litre of the standard medium with 65 g/l glucose. The pH was maintained at 5.8. **(b)** Time course of solka floc degradation by *Clostridium* H.10 co-cultured with *C. acetobutylicum* NCIB 619. The experiment was performed in a fermentor that contained 1 litre of the standard medium with 3% solka floc. *Clostridium* H.10 was inoculated for 48 h at pH 7.0, then co-culture was initiated with *C. acetobutylicum* and the pH maintained at 5.8. Reprinted from Ford *et al.* (1983), with permission. Copyright John Wiley and Sons, Inc.

grown in isolated culture on glucose, cellobiose or xylose, it produces little or no acetic acid in co-culture with *Z. anaerobia*. Co-culturing of these microorganisms resulted in both a suppression of acetic acid production and higher yields of ethanol (see *Table 10*).

Another possibility for enhancing ethanol production involves the expression of cellulase genes from fungi such as *T. reesei* into efficient ethanol producers such as yeast. This would confer on the yeast the ability to utilize cellulose directly. Shoemaker and co-workers (Shoemaker *et al.*, 1983; Shoemaker, 1984) have cloned the genes for cellobiohydrolase I and endoglucanase I from *T. reesei* strain L27 and introduced them in recombinant form into yeast (*S. cerevisiae*). Protein secreted by the transformed yeast cells contained both enzymes. Expression of the *Aspergillus* β-glucosidase genes in yeast would also be necessary to complete the biodegradation of cellulose. It will be interesting to see whether such transformed yeast are capable of fermenting cellulose to ethanol.

The work of Ford *et al.*, provides an example of cellulose fermentation by a co-culture of a mesophilic cellulolytic *Clostridium* (strain H.10) and *Clostridium acetobutylicum* (Ford *et al.*, 1983). *Clostridium* H.10 in monoculture degrades cellulose to produce acetic acid and ethanol (*Table 9*). *C. acetobutylicum* produces acetic and butyric acids in the initial stages of a fermentation based on glucose utilization, but in the later stages, the acids are re-assimilated to butanol and acetone (*Figure 7a*). In co-culture, however, a different pattern of excreted metabolites was found (*Figure 7b*), with butyric acid being the major product. Apparently, for a solvent-type fermentation to occur, a high rate of glucose feeding is required, which is not provided during solka floc degradation by *Clostridium* H.10 in co-culture with *C. acetobutylicum*.

Therefore, for some fermentations (as in the case of butanol production), it is desirable to utilize glucose rather than cellulose. Under these circumstances, the cellulose-derived glucose should be used as the substrate. Furthermore, anaerobic bacteria can only ferment biomass that has been pre-treated to remove lignin. The advantage of direct cellulose utilization is that enzyme production and recovery are not required. For high yields of a desired product, indirect utilization of cellulose may be preferable.

At Oak Ridge National Laboratory (ORNL), the anaerobic digestion of cellulosic wastes (blotting paper, paper towels, laboratory coats) has been carried out using sludge from the municipal treatment plant in Oak Ridge, Tennessee. Since sludge contains a mixture of acidogenic and methanogenic bacteria, such a process provides methane gas as a result of the direct utilization of cellulose. The main purpose of such a digester is to reduce the volume of radioactive cellulosic wastes at ORNL. The initial volume of such wastes is reduced by $90-95\%$, which simplifies disposal (Lee and Donaldson, 1984).

There are also several species of anaerobic rumen bacteria that will digest insoluble cellulose (for a review, see Lin *et al.*, 1985). Some of the fermentation products of the major cellulolytic rumen microorganisms in pure and co-culture are shown in *Table 11*.

Concluding remarks

There are many problems still to be overcome in the utilization of cellulose in fermentation processes on a practical scale. Under some circumstances, it may be preferable to hydrolyse the cellulose to glucose prior to fermentation. The problems

Table 11. Fermentation products of the major cellulolytic rumen microorganisms in pure culture and co-culture.

Bacterial species	Products (mol/mol glucose)								
	Substrate	Acetate	Propionate	Succinate	Ethanol	Lactate	H_2^a	CH_4	CO_2
Ruminococcus albus	Glucose	0.74				0.69		2.37		1.43
Ruminococcus albus + *Wolinella succinogenes*	Glucose	1.47				0.0	3.84	3.84		1.47
Ruminococcus flavefaciens	Cellulose	1.07		0.93				0.94	0.0	-0.48^b
Ruminococcus flavefaciens + *Methanobacterium ruminantium*	Cellulose	1.89		0.11				0.0	0.83	0.94
Anaerobic fungi	Cellulose	0.73			0.37	0.67	1.18	0.0	0.38
Anaerobic fungi + methanogen	Cellulose	1.35		1.38	0.19	0.03	0.02	0.50	0.89
Bacteroides succinogenes	Cellulose	0.85	0.0	1.38					
Bacteroides succinogenes + *Selenomonas ruminantium*	Cellulose	0.87	1.38	0.0					

[a] H_2 equivalents include formate for the anaerobic fungi and electrons used to reduce fumarate to succinate by *Wolinella succinogenes*.
[b] Negative sign indicates CO_2 uptake.
Reproduced from Lin *et al.* (1985), with permission. Copyright Butterworth and Co. (Publishers), Ltd.

associated with acid and enzymatic hydrolysis of cellulose have been addressed in this report, as well as those involved with direct cellulose utilization. Continued financial support of fundamental research in these areas by governments and industry is essential if these problems are to be solved.

Acknowledgements

The author would like to thank Ms D.Weaver for secretarial assistance. The preparation of this manuscript was supported in part by the Office of Basic Energy Sciences, US Department of Energy, under contract No. DE-AC05-84OR21400 with Martin Marietta Energy Systems, Inc.

References

Amrhein,N., Frank,G., Lemm,G. and Luhmann,H.-B. (1983) Inhibition of lignin formation by L-α-amino-oxy-β-phenylpropionic acid, an inhibitor of phenylalanine ammonia-lyase. *Eur. J. Cell Biol.*, **29**, 139−144.

Amrhein,N. and Gödeke,K.-H. (1977) α-Aminooxy-β-phenylpropionic acid − a potent inhibitor of L-phenyl-alanine ammonia-lyase *in vitro* and *in vivo*. *Plant Sci. Lett.*, **8**, 313−317.

Asther,M. and Khan,A.W. (1984) Influence of the presence of *Zymomonas anaerobia* on the conversion of cellobiose, glucose, and xylose to ethanol by *Clostridium saccharolyticum*. *Biotechnol. Bioeng.*, **26**, 970−972.

Bungay,H.R. (1984) Status of cellulose hydrolysis processes. *Interciencia*, **9**, 87−90.

Churilova,I.V., Maksimov,V.I. and Klesov,A.A. (1980) Cellobiose as a regulator of the activity of endo-glucanases of cellulase complexes. Mechanism of regulation. *Biokhimiya*, **44**, 1663−1665.

Cooney,C.L., Wang,D.I.C., Wang,S.-D., Gordon,J. and Jiminez,M. (1978) Simultaneous cellulose hydrolysis and ethanol production by a cellulolytic anerobic bacterium. *Biotechnol. Bioeng. Symp. Ser.*, **8**, 103−114.

Dale,B.E., Henk,L.L. and Shiang,M. (1985) Fermentation of lignocellulosic materials treated by ammonia freeze-explosion. *Dev. Ind. Microbiol.*, **26**, 223−233.

Deeble,M.F. and Lee,J.M. (1985) Enzymatic hydrolysis of cellulosic substances in an attrition bioreactor. *Biotechnol. Bioeng. Symp. Ser.*, **15**, in press.

Deshpande,M.V. and Eriksson,K.-E. (1984) Reutilisation of enzymes for saccharification of lignocellulosic materials. *Enzyme Microb. Technol.*, **6**, 338−340.

Fadda,M.B., Dessi,M.R., Maurici,R., Rinaldi,A. and Satta,G. (1984) High efficient solubilization of natural lignocellulosic materials by a commercial cellulase immobilised on various solid supports. *Appl. Microbiol. Biotechnol.*, **19**, 306−311.

Ford,O., Petitdemange,E., Petitdemange,H. and Engasser,J.-M. (1983) Cellulase fermentation by a coculture of a mesophilic cellulolytic *Clostridium* and *Clostridium acetobutylicum*. *Biotechnol. Bioeng. Symp. Ser.*, **13**, 217−224.

Gould,J.M. (1984) Alkaline peroxide delignification of agricultural residues to enhance enzymatic saccharification. *Biotechnol. Bioeng.*, **26**, 46−52.

Gould,J.M., Kerley,M.S., Fahey,G.C.,Jr. and Berger,L.L. (1985) Alkaline hydrogen peroxide treatment to enhance utilisation of lignocellulose. Paper presented at the 190th Annual Meeting of the American Chemical Society, September 8−13, Chicago, Illinois.

Grethlein,H.E. (1985) The effect of pore-size distribution on the rate of enzymatic hydrolysis of cellulosic substrates. *Bio/Technology*, **3**, 155−160.

Grous,W., Converse,A., Grethlein,H. and Lynd,L. (1985) Kinetics of cellobiose hydrolysis using cellobiase composites from *Trichoderma reesei* and *Aspergillus niger*. *Biotechnol. Bioeng.*, **27**, 463−470.

Hamilton,J.J., Dale,B.E., Ladisch,M.R. and Tsao,G.T. (1984) Effect of ferric tartrate/sodium hydroxide solvent pretreatment on enzyme hydrolysis of cellulose in corn residue. *Biotechnol. Bioeng.*, **26**, 781−787.

Hawkins,A.F., Owen,T.R., Morley,J.S. and Hayward,C.F. (1981) Method of improving or maintaining digestibility of fodder crops. US patent No. 4,304,592.

Karube,I., Tanaka,S., Shirai,T. and Suzuki,S. (1977) Hydrolysis of cellulase-bead fluidised bed reactor. *Biotechnol. Bioeng.*, **19**, 1183−1191.

Kirk,T.K., Croan,S., Tien,M., Murtagh,K.E. and Farrell,R.L. (1986) Production of multiple ligninases by *Phanerochaete chrysosporium*: effect of selected growth conditions and use of a mutant strain. *Enzyme Microb. Technol.*, **8**, 27−32.

Klyosov,A.A. (1985) Adsorption of cellulases on cellulose and lignin. Paper presented at the Symposium on the Pretreatment of Lignocellulosic Materials, Forest Research Institute, Rotorua, New Zealand.

Kuwahara,M., Glenn,J.K., Morgan,M.A. and Gold,M.H. (1984) Separation and characterisation of two H_2O_2-dependent oxidases from ligninolytic cultures of *Phanerochaete chrysosporium. FEBS Lett.*, **169**, 247−250.

Ladisch,M.R. and Tsao,G.T. (1986) Engineering and economics of cellulose saccharification systems. *Enzyme Microb. Technol.*, **8**, 66−69.

Lee,D.D. and Donaldson,T.L. (1984) Dynamic simulation model for anerobic digestion of cellulose. *Biotechnol. Bioeng. Symp. Ser.*, **14**, 503−508.

Lee,J.M. and Woodward,J. (1983) Properties and application of immobilised β-D-glucosidase coentrapped with *Zymomonas mobilis* in calcium alginate. *Biotechnol. Bioeng.*, **25**, 2441−2451.

Lin,K.W., Patterson,J.A. and Ladisch,M.R. (1985) Anaerobic fermentation: microbes from ruminants. *Enzyme Microb. Technol.*, **7**, 98−106.

Mandels,M. (1985) Application of cellulases. *Biochem. Soc. Trans.*, **13**, 414−416.

Mandels,M., Medeiros,J.E., Andreotti,R.E. and Bissett,F.H. (1981) Enzymatic hydrolysis of cellulose: evaluation of cellulase culture filtrates under use conditions. *Biotechnol. Bioeng.*, **23**, 2009−2026.

Montenecourt,B.S. (1983) *Trichoderma reesei* cellulases. *Trends Biotechnol.*, **1**, 156−161.

Nystrom,J.M., Greenwald,C.G., Harrison,F.G. and Gibson,E.D. (1984) Making ethanol from cellulosics. *Chem. Eng. Prog.*, **80**, 68−74.

Ohlson,I., Trädgardh,G. and Hahn-Hägerdal,B. (1983) Recirculation of cellulolytic enzymes in an ultrafiltration membrane reactor. *Acta Chem. Scand.*, **B37**, 737−738.

Puri,V.P. (1984) Effect of crystallinity and degree of polymerisation of cellulose on enzymatic saccharification. *Biotechnol. Bioeng.*, **26**, 1219−1222.

Reese,E.T. (1982a) Elution of cellulase from cellulose. *Proc. Biochem.*, **17**, 2−8.

Reese,E.T. (1982b) Protection of *Trichoderma reesei* cellulases from inactivation due to shaking. In *Solution Behaviour of Surfactants*. Mittal,K.L. and Fendler,E.J. (eds), Plenum Publishing Corporation, New York, Vol. 2, pp. 1487−1504.

Rogalski,J., Szczodrak,J., Dawidowicz,A., Ilczuk,Z. and Leonowicz,A. (1985) Immobilisation of cellulase and D-xylanase complexes from *Aspergillus terreus* F-413 on controlled porosity glasses. *Enzyme Microb. Technol.*, **7**, 395−400.

Rogers,P.L., Lee,K.J. and Tribe,D.E. (1980) High productivity ethanol fermentations with *Zymomonas mobilis. Proc. Biochem.*, **15**, 7−11.

Ryu,S.K. and Lee,J.M. (1983) Bioconversion of waste cellulose by using an attrition bioreactor. *Biotechnol. Bioeng.*, **25**, 53−65.

Shoemaker,S.P. (1984) The cellulase system of *Trichoderma reesei: Trichoderma* strain improvement and expression of *Trichoderma* cellulases in yeast. Paper presented at Biotech '84 USA: Online Publications, Pinner, UK.

Shoemaker,S.P., Raymond,J.C. and Bruner,R. (1981) Cellulases: diversity amongst improved *Trichoderma* strains. In *Trends in the Biology of Fermentations for Fuels and Chemicals*. Hollaender,A., Rabson,R., Rogers,P., San Pietro,A., Valentine,R. and Wolfe,R., (eds), Plenum Press, New York, pp. 89−109.

Shoemaker,S., Schweickart,V., Ladner,M., Gelfand,D., Kwok,S., Myambo,K. and Innis,M. (1983) Molecular cloning of exocellobiohydrolase I derived from *Trichoderma reesei* strain L27. *Bio/Technology*, **1**, 691−696.

Sinitsyn,A.P., Bungay,M.L., Clesceri,L.S. and Bungay,H.R. (1983) Recovery of enzymes from the insoluble residue of hydrolysed wood. *Appl. Biochem. Biotechnol.*, **8**, 25−29.

Sinitsyn,A.P., Clesceri,L.S. and Bungay,H.R. (1982) Inhibition of cellulases by impurities in steam-exploded wood. *Appl. Biochem. Biotechnol.*, **7**, 455−458.

Spangler,D.J. and Emert,G.H. (1986) Simultaneous saccharification/fermentation with *Zymomonas mobilis. Biotechnol. Bioeng.*, **28**, 115−118.

Sundstrom,D.W., Klei,H.E., Coughlin,R.W., Biederman,G.J. and Brouwer,C.A. (1981) Enzymatic hydrolysis of cellulose to glucose using immobilised β-glucosidase. *Biotechnol. Bioeng.*, **23**, 473−485.

Takagi,M. (1984) Inhibition of cellulase by fermentation products. *Biotechnol. Bioeng.*, **26**, 1506−1507.

Taya,M., Suzuki,Y. and Kobayashi,T. (1984) A thermophilic anaerobe (*Clostridium* sp.) utilising various biomass-derived carbohydrates. *J. Ferment. Technol.*, **62**, 229−239.

Tien,M. and Kirk,T.K. (1983) Lignin-degrading enzyme from the hymenomycete *Phanerochaete chrysosporium* Burds. *Science*, **221**, 661−663.

Tjerneld,F., Persson,I., Albertsson,P.-Å. and Hahn-Hägerdal,B. (1985) Enzymatic hydrolysis of cellulose in aqueous two-phase systems. 1. Partition of cellulases from *Trichoderma reesei. Biotechnol. Bioeng.*, **27**, 1036−1043.

Vallander,L. and Eriksson,K.-E. (1985) Enzymic saccharification of pretreated wheat straw. *Biotechnol. Bioeng.*, **27**, 650−659.

Volfová,O., Suchardová,O., Panós,J. and Krumphanzl,V. (1985) Ethanol formation from cellulose by thermophilic bacteria. *Appl. Microbiol. Biotechnol.*, **22**, 246−248.

Wood,T.M. (1985) Properties of cellulolytic enzyme systems. *Biochem. Soc. Trans.*, **13**, 407−410.

Woodward,J. and Arnold,S.L. (1981) The inhibition of β-glucosidase activity in *Trichoderma reesei* C30 cellulase by derivatives and isomers of glucose. *Biotechnol. Bioeng.*, **23**, 1553−1562.

Woodward,J., Baird,A.R. and Parrish,C. (1986) Development of a method for the recovery of the enzyme cellulase from aqueous solution. In *Proceedings of World Congress III of Chemical Engineering*. Tokyo, Japan, in press.

Woodward,J., Whaley,K.S., Zachry,G.S. and Wohlpart,D.L. (1981) Thermal stability of *Trichoderma reesei* C30 cellulase and *Aspergillus niger* β-glucosidase after pH and chemical modification. *Biotechnol. Bioeng. Symp. Ser.*, **11**, 619−629.

Woodward,J. and Wiseman,A. (1982) Fungal and other β-D-glucosidases − their properties and application. *Enzyme Microb. Technol.*, **4**, 73−79.

Woodward,J. and Zachry,G.S. (1982) Immobilisation of cellulase through its carbohydrate side chains − a rationale for its recovery and reuse. *Enzyme Microb. Technol.*, **4**, 245−248.

CHAPTER 5

Transport of carbohydrates by bacteria

P.J.F.HENDERSON and M.C.J.MAIDEN

Department of Biochemistry, University of Cambridge, Tennis Court Road, Cambridge CB2 1QW, UK

Introduction

Bacteria often inhabit environments where nutrients are in short supply, and different species must compete with each other for the available metabolites. Accordingly, they expend metabolic energy in order to sequester sugars and other essential nutrients (K^+, NH_4^+, P_i, vitamins, etc.) and to achieve intracellular concentrations sufficient for optimal growth rates. This expenditure can amount to $20-30\%$ of an organism's 'energy budget', when a carbohydrate is fermented under anaerobic conditions to yield only $2-3$ mol ATP per mol sugar (Muir *et al.*, 1985; Cooper, 1986; Dawes, 1986). Since the efficiencies of the transport steps may therefore influence cell yield and growth rate (Koch, 1971; Button, 1985; Muir *et al.*, 1985) an understanding of the transport processes is important from the viewpoints of both the academic researcher seeking to understand bacterial cell physiology, and the production manager trying to maintain the profitability of a fermentation process.

The major part of this chapter is a review of the current understanding of the four mechanisms by which bacteria are known to energize the transport of sugars into their cells:

(i)   the phosphotransferase systems [utilizing phosphoenolpyruvate (PEP) as energy source, *Figure 1*];

(ii) the binding protein systems (utilizing ATP − or a closely related glycolytic product − as energy source, *Figure 1*);

(iii) The H^+−sugar coupled ('symport') systems utilizing the trans-membrane electrochemical gradient of protons generated by respiration or ATPase (*Figure 1*); and

(iv) the Na^+−sugar coupled systems utilizing the trans-membrane gradient of Na^+ ions (*Figure 2*).

The reader will be referred to more specialized reviews for detailed information on each type.

In the last part the experimental techniques currently available for the investigation and manipulation of transport processes are illustrated by a study of an arabinose

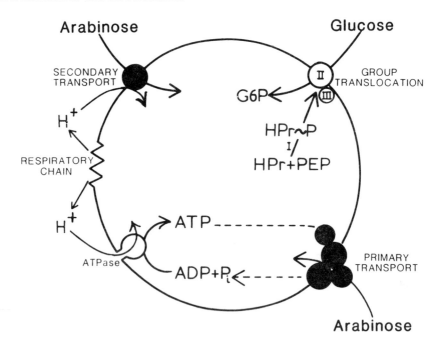

Figure 1. Scheme for energization of arabinose and glucose transport in *E. coli*. The large circle represents the cytoplasmic membrane. The H^+-arabinose symporter (AraE, top left, see *Table 1*) is energized by the proton-motive force, which is generated by respiration and also used for ATP synthesis. The binding protein arabinose transport system (AraF, bottom right, see *Table 1*) is energized by a phosphorylated compound, probably ATP. Both systems transport chemically unmodified arabinose. This contrasts with the glucose phosphotransferase (top right), which uses PEP to phosphorylate the sugar via the HPr protein, Enzyme I, Enzyme III^{glc} and Enzyme II (see text). So far there is evidence for at least two protein components of the AraF system, but four are drawn because this would be consistent with the number present in other such transport systems (see text).

transport system of *Escherichia coli*. The combination of genetical and biochemical approaches avoids some of the problems of studying such hydrophobic proteins of low abundance. A routine strategy may be summarized as follows.

(i)   Physiological techniques are used to identify a novel transport system, which is then characterized in terms of its specificity, kinetics, energization and regulation.

(ii) Genetic manipulations are used to mutate, identify and clone the new gene(s) coding for the transport system.

(iii) The DNA sequencing of the gene establishes the primary structure of the transport protein(s).

(iv) The protein(s) are identified and purified, if possible using an over-expression vector.

The next, more challenging, stages are to establish the three-dimensional structures of the proteins and to deduce their molecular mechanisms. The current state of technology already permits the experimenter to manipulate the type of transport process (and metabolic pathway) available for carbohydrate utilization in a microorganism; such manipulation is of value in biotechnology.

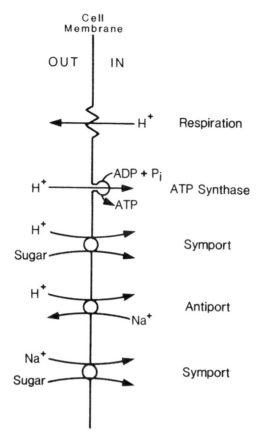

Figure 2. A scheme for the operation of proton- and sodium-coupled transport systems in the membrane of *E. coli*.

Most is known of the carbohydrate transport processes in the Gram-negative organisms, *Escherichia coli* and *Salmonella typhimurium*, because of their susceptibilty to sophisticated genetic techniques. Nevertheless, some of their transport mechanisms occur in many other microorganisms and even, as we have recently discovered (see below), in man himself.

Penetration of the cell wall by carbohydrates

The cell walls of Gram-negative bacteria have a complex multi-layered structure (Rogers *et al.*, 1980). In *E. coli* and a number of other species the evidence suggests that monosaccharides, disaccharides and probably trisaccharides, penetrate these layers (lipopolysaccharide, outer membrane, peptidoglycan, periplasm) at rates that do not limit cell growth (Nikaido and Vaara, 1985). This is achieved by at least three factors. Firstly, the lipopolysaccharide layer is permeable to sugars, though it may be impermeable to more hydrophobic molecules including antibiotics (Nikaido and Vaara, 1985; Nikaido, 1986). Secondly, the outer membrane contains channel-forming trimeric proteins ('porins'), acting as molecular sieves that permit simple diffusion of solutes of molecular weights up to 900, including di- and trisaccharides (di Renzio *et al.*, 1978; Nikaido and Vaara, 1985; Nikaido, 1986; Nakai, 1986). Thirdly, the presence in the outer membrane of porin-like proteins which exhibit some specificity for the permeant molecule, and pass the substrate (we presume) to high affinity binding proteins in the periplasm (Nikaido and Vaara, 1985; Furlong, 1986).

An elegant series of biophysical measurements has culminated in a three-dimensional picture of porin structure, in which three channels on the outer face of the membrane merge into a single channel at the inner face (Engel *et al.*, 1985). These proteins contain a high proportion of anti-parallel β-pleated sheet structure (Kleffel *et al.*, 1985).

If a substrate of trisaccharide size or greater is to be utilized effectively for bacterial growth then a relatively specific porin may be necessary to optimize its penetration of the outer membrane. One example is the transport of maltodextrins through the LamB protein (Hengge and Boos, 1983). *E. coli* mutants with defective LamB protein in the outer membrane are only mildly impaired in the active transport of maltose (at concentrations below 10 μM) but are progressively more impaired in the utilization of higher oligosaccharides of the maltodextrin series (Hengge and Boos, 1983); loss of the MalE periplasmic binding protein, however, abolishes transport of all of these sugars (Hengge and Boos, 1983).

Penetration of the cell membrane by carbohydrates

The inner cell membrane, a protein-containing phospholipid bilayer (Rogers *et al.*, 1980), is the barrier preventing the entry of most ambient solutes into the bacterial cell. Nutrient uptake is therefore effected by integral membrane transport proteins, either singly or in complexes, the majority of which are synthesized only in the presence of their substrate (see below). Energization of transport is effected at this inner membrane. Amongst its many other functions are the processes of respiration, ATP synthesis, maintenance of the K^+ gradient and osmoregulation. These are themselves transport processes (Mitchell, 1963, 1966; West and Mitchell, 1974; Epstein, 1982; Cairney *et al.*, 1986). So we can visualize the occurrence of permanent and 'plug-in' transport modules in the membrane, some of which are dependent on others (*Figures 1* and *2*). For example, only one inducible protein is required for lactose transport (Kaback, 1986), but the energization of its accumulation requires the respiratory chain or ATPase activity (see *Figures 1* and *2*), which are more permanent features of the membrane (Ingledew and Poole, 1984; Walker *et al.*, 1984; Futai and Kanazawa, 1983).

Before considering the individual sugar transport systems it is useful to define some terms often used in their characterization. A good introduction is the book by West (1983).

Diffusion

Diffusion is non-saturable, non-energized transport through a 'pore' (or directly through the lipid) in which

$$v = P A c \qquad\qquad \text{Equation 1}$$

where v = velocity, P = permeability coefficient for the particular solute, A = area and c = difference in solute concentration across the cell membrane. Diffusion has a low temperature coefficient (v $\propto$ absolute temperature) and is unspecific. Transport through porins is of this type. Due to differences in their molecular size, the permeability coefficients for the different sugars vary over a 100-fold range from pentoses (L-arabinose) to disaccharides (lactose) for the porin OmpF (Nikaido and Rosenberg, 1983) without impairing utilization for growth.

Facilitated diffusion

This is saturable, non-energized transport catalysed, in cells, by a protein. The Michaelis–Menten relationship (see Henderson, 1985, for a review)

$$v = \frac{V_{max} \, [S]}{K_m + [S]} \qquad \text{Equation 2}$$

(V_{max} = maximum velocity, K_m = [S] where v is $V_{max}/2$) often adequately relates the initial rate of transport (v) to initial substrate concentration ([S] = c at zero time). As with enzyme reactions, there is a high temperature coefficient and, usually, strong substrate specificity. This mechanism depends on a concentration gradient across the cell wall being maintained by the metabolism of the substrate.

Glycerol transport is the only known example of facilitated diffusion in *E. coli* (Heller *et al.*, 1980; Lin, 1986).

Active transport

The term active transport is used to describe the net transport of a solute across a biological membrane from a low electrochemical potential to a higher potential (see West, 1983). It shows the following characteristics.

(i)   Accumulation occurs against a concentration gradient.
(ii)   The solute is not chemically modified during translocation.
(iii)   Saturable steady state kinetics are observed, often following the Michaelis–Menten relationship (above).
(iv)   There is a high temperature coefficient typical of enzyme-catalysed reactions.
(v)   Substrate specificity is restricted.
(vi)   There is input of metabolic energy.

The energization is achieved in at least three ways for active transport of sugars in *E. coli*. One is the direct energization of transport of maltose, galactose, arabinose, xylose and ribose by a phosphorylated product of glycolysis, probably ATP (*Figure 1*; Furlong, 1986). A second is the linkage of lactose, galactose, arabinose, xylose, fucose or rhamnose transport to the electrochemical gradient of protons established by respiration or ATP hydrolysis (*Figures 1, 2*; West and Mitchell, 1972; Kaback, 1986; Henderson, 1986). A third is the linkage of melibiose transport to the gradient of Na^+ ions (*Figure 2*, Stock and Roseman, 1971; Tsuchiya *et al.*, 1977), but this is perhaps just a special case of the second mechanism.

In *E. coli* the trans-membrane H^+ gradient would appear to be the 'common currency' of many energized transport reactions, and the Na^+ gradient of relatively few. However, in other organisms living in salt environments the Na^+ gradient is the dominant factor maintained by a primary Na^+ pump (Tokuda, 1986; Dimroth, 1986; Buckel, 1986), as it is in multicellular eukaryotes.

Group translocations

This type of system catalyses the transport and concomitant chemical modification of the transported solute. For a range of carbohydrates in many species of bacteria (*Table 1*), PEP is the donor to produce internal sugar-phosphate from external free sugar (Saier, 1985; Postma and Lengeler, 1985).

Primary and secondary transport, symport and antiport

The transport processes outlined above embrace a variety of molecular mechanisms. It is conceptually helpful to classify them further into 'primary' and 'secondary' mechanisms (reviewed by Rosen and Kashket, 1978). Secondary transport can be subdivided into 'symport' or 'antiport', terms introduced by Mitchell (1963, 1973).

Primary transport involves the direct conversion of chemical or photosynthetic energy into an electrochemical potential of transport substrate across the membrane barrier. Thus, translocation of protons driven by oxidation of respiratory substrates (Mitchell, 1966; Ingledew and Poole, 1984), by hydrolysis of ATP (Mitchell, 1966; Kagawa, 1978; Futai and Kanazawa, 1983) or by light energy absorbed by bacteriorhodopsin (Stoeckenius *et al.*, 1979; Henderson and Unwin, 1975) all fall into this class. The phosphotransferase mechanism of carbohydrate transport, energized by PEP, is primary transport. It seems that transport systems involving binding proteins are also of the primary type, but energized by ATP (see below).

Secondary transport involves the conversion of a pre-existing electrochemical gradient, usually of H^+ or Na^+ ions, into a new electrochemical gradient of the transported species. Thus the ultimate energy source for secondary transport systems is a primary chemical or photochemical conversion. For example, in *E. coli* primary proton ejection by respiration or ATPase powers secondary $H^+ -$ sugar *symport* (obligatory coupling of H^+ and solute movement in the *same* direction; Mitchell, 1973; *Figure 1*) or secondary $H^+ - Na^+$ *antiport* (the obligatory coupling of H^+ and solute movement in the *opposite* direction; Mitchell, 1973, *Figure 2*). The resulting Na^+ gradient can be further coupled to melibiose transport by a $Na^+ -$ melibiose symport, so that net melibiose accumulation is driven by respiration (or ATPase) via H^+ and Na^+ gradients (*Figure 2*).

Why do bacteria have more than one transporter per solute?

A catalogue of the substrate specificities of individual transport systems (*Table 1*), shows that in *E. coli* some sugars may enter the cell by more than one transport system. Glucose is a potential substrate of the PtsG, PtsM, PtsF, BglC, ScrA, GalP and Mgl systems encompassing three different mechanisms of energization (*Table 1*). Galactose is a potential substrate of GalP, Mgl, PtsG, LacY, AraE and AraF (Kornberg, 1976), again encompassing three different mechanisms.

In the laboratory, these options for carbohydrate transport are usually limited by inducer specificity. In the presence of glucose PtsG is the only route for glucose entry, since other transport systems are not only uninduced, but also their expression is repressed by glucose (Postma and Lengeler, 1985; Saier, 1985). GalP and Mgl are the physiological routes for galactose entry, since only these two are induced by galactose (Wilson, 1974; Daruwalla *et al.*, 1981). Similarly, AraE and AraF are two physiological

Table 1. Sugar transport systems of *E. coli*.

Sugar	Gene and location (min)	Other substrates
Phosphotransferase systems		
Glucose	*ptsG* (24)	Mannose, L-sorbose, glucosamine, methyl-α-glucoside, 5-thioglucose
Mannose	*ptsM* (40)	Glucose, N-acetylglucosamine, glucosamine, fructose, 2-deoxyglucose
Fructose	*ptsF* (47)	Glucose, L-sorbose, mannose, xylitol
N-Acetylglucosamine	*nagE* (16)	Streptozotocine
Glucitol	*gutA* (58)	Galactitol, mannitol, arabinitol, fructose, glucose, 2-deoxyarabinohexitol
Mannitol	*mtlA* (81)	Glucitol, arabinitol, 2-deoxyarabinitol
Galactitol	*gatA* (46)	Glucitol, 2-deoxygalactitol, arabinitol
Trehalose	*tre* (26)	
Sorbose	*sorA* (91)	L-Sorbose
β-Glucosides	*bglC* (83)	Glucose, cellobiose, arbutin, salicin, O-nitrophenyl-β-glucoside
Sucrose	*scrA*	Glucose
Binding protein transport systems		
Maltose	*malB* (91)	Maltodextrins up to maltohexose, methyl-β-maltoside, 5-thiomaltose
Galactose	*mgl* (45)	Glucose, fucose, methyl-β-galactoside, glyceryl-β-galactoside
Arabinose	*araF* (45)	Fucose
Xylose	*xylF* (79)	
Ribose	*rbs* (84)	
Cation-linked transport systems		
Lactose	*lacY* (8)	Galactose, many β-galactosides
Melibiose	*melB* (93)	Galactinol, β-galactosides
Galactose	*galP* (64)	Glucose, fucose, talose, 2-deoxygalactose, 6-fluoro-galactose, 4-fluorogalactose, 6-fluoroglucose, 6-deoxyglucose, 2-deoxyglucose
Arabinose	*araE* (61)	Fucose
Xylose	*xylE* (91)	6-Deoxyglucose
L-Rhamnose	*rha* (87)	L-Mannose, L-lyxose
L-Fucose	[*fuc* (60)]	L-Galactose, D-arabinose

The modified data are taken from Postma and Lengeler (1986), Henderson (1986) and Henderson and Macpherson (1986). Sugar names imply the D-form except for L-arabinose and those indicated.

routes for arabinose entry (Daruwalla *et al.*, 1981) and XylE and XylF for xylose entry (Davis *et al.*, 1984; Davis, 1985).

The existence of different types of transport process for one carbohydrate can be rationalized on the basis of their biochemistry. Galactose, arabinose and xylose are each transported by an H^+−sugar symporter (GalP, AraE and XylE, respectively; Henderson and Macpherson, 1986) and a binding protein-dependent system (Mgl, AraF and XylF; Furlong, 1986). The H^+ symporters have relatively high K_m values (50−450 μM) and may be termed 'low-affinity', whereas the binding protein systems are 'high-affinity' with K_m values of 0.2−6 μM (Henderson, 1986). Their V_{max} values are similar. The H^+ symporters are probably more efficient in terms of energy expenditure per molecule transported (Henderson, 1980; Muir *et al.*, 1985; see below), but they achieve lower concentration gradients ($[S]_{inside}/[S]_{outside}$) than the high-affinity systems (Hengge and Boos, 1983). It is likely that the H^+ symport mechanism permits energetically more efficient, faster, growth in conditions of carbohydrate plenty, whereas the binding protein mechanism enables scavenging in conditions of low carbohydrate concentration. High-affinity systems will presumably enable an organism to out-compete others with only low-affinity systems under conditions where their energy cost is not a determining factor. The binding proteins also contribute to chemotactic responses to certain solutes, as does the Enzyme II of the phosphotransferase (Adler, 1975; Saier, 1985).

The biotechnologist will probably require an organism used in a production process to have an energy-efficient transport system of the low-affinity type (sugars are usually provided at high concentrations) to allow rapid growth with a high growth yield. It is now possible to manipulate transport systems genetically to achieve this, if a suitable transport system is not already present in the organism (see below).

The next sections provide an outline of the substrate specificities, proteins, genes, regulatory mechanisms, energetics and distinguishing features of each type of transport system.

The phosphotransferase sugar transport systems

This mechanism was first characterized by Kundig (1974), and the extensive subsequent studies were recently reviewed by Postma and Lengeler (1985) and Saier (1985). The major sugar transported by this route is glucose, a common carbohydrate component in feedstocks for industrial fermentations (in starch, maltose or sucrose as well as glucose itself).

Sugar substrates

The sugar substrates that enter *E. coli* by a phosphotransferase mechanism are listed in *Table 1* (see also Table 2 of Postma and Lengeler, 1985). The dominating sugar is glucose, because its presence depresses the utilization of other carbohydrates, possibly because its entry into metabolism is particularly economical of cellular energy reserves (Muir *et al.*, 1985). It is quite feasible that sugars not utilizing a phosphotransferase system in *E. coli* nevertheless do so in another organism, for example lactose in Gram-positive bacteria (Saier, 1985) and galactose in *Streptococci* (Thompson, 1980). The phosphotransferase mechanism appears to be widely disseminated amongst bacterial species (Saier, 1985), but it has not been reported to occur in other organisms to our knowledge.

Proteins and genes

In the cytoplasm a soluble protein HPr (mol. wt 9000, gene location 52 min) is phosphorylated at the N-1 position of His_{15}. This reaction is catalysed by 'Enzyme I' (mol. wt 68 000, gene location 52 min), which is itself reversibly phosphorylated at a His residue by PEP (*Figure 1*, Postma and Lengeler, 1985). The HPr-phosphate is utilized by the membrane-bound 'Enzyme II' to phosphorylate the sugar during transport, either directly or via 'Enzyme III' (mol. wt 18 500, gene location 52 min for Enzyme III^{glc}). Enzyme I and HPr are components common to most phosphotransferase systems, but there is an Enzyme II specific for each sugar, the gene location of each being scattered around the chromosome (*Table 1*). An exception is the transport system for fructose, which has a separate 'FPr' or 'pseudo HPr' (Postma and Lengeler, 1985; Saier, 1985; Kornberg, 1986) possibly coupled with a fructose-specific Enzyme III^{Fru} (Waygood *et al.*, 1984).

The primary structure of the Enzyme II for β-glucosides, deduced from the DNA sequence of *bglC*, has some homology with the mannitol Enzyme II (Lee and Saier, 1983; Bramley and Kornberg, 1987) and, to a lesser extent, HPr (unpublished).

Regulation

In general, the expression of genes for Enzyme I and HPr is de-repressed in *E. coli* whatever the growth conditions (Postma and Lengeler, 1985). However, the majority of genes for Enzymes II are not well expressed unless the particular sugar substrate is present in the environment (Saier, 1985). It is not often known whether the inducer is external sugar, internal sugar − phosphate, or some other metabolite, and the control mechanisms await elucidation. The expression of Enzyme II^{glc} is usually significantly de-repressed under all growth conditions, though in individual strains the level may well be increased by glucose (Postma and Lengeler, 1985).

Thus, phosphotransferase transport systems are generally 'inducible'. However, this induction is likely to be depressed if glucose is also present (Saier, 1985; Postma and Lengeler, 1985). The depression at the transcriptional level may be explained by the ability of the glucose phosphotransferase to deplete intracellular cAMP levels so that *crp* protein no longer promotes the expression of the genes concerned (Saier, 1985). This mechanism is loosely described as catabolite repression, and it is probably mediated through Enzyme III^{glc} (Saier, 1985; Postma and Lengeler, 1985). The presence of glucose may also exclude or expel the inducer of other phosphotransferase systems (Saier, 1985). Additionally, the activities of phosphotransferase proteins may be modulated directly by the phosphorylation state of a serine residue in HPr, inhibition of Enzymes II by carbohydrate phosphates, competition for HPr phosphate and sugar substrates, and by the membrane potential; these and other possible regulatory features are discussed in detail elsewhere (Postma and Lengeler, 1985; Saier, 1985).

Most importantly, the presence of glucose represses expression, and Enzyme III^{glu} may inhibit directly transport systems (and metabolic pathways) that are not of the phosphotransferase type (Saier, 1985; de Boer *et al.*, 1986; and see below).

Energetics of phosphotransferase transport

Obviously, one molecule of PEP is used per molecule of sugar transported by this mechanism, reducing the yield of ATP from metabolism of the sugar by one. However, in the case of glucose it arrives in the cytoplasm as glucose 6-phosphate, ready for glycolysis without further modification or energy input (Muir *et al.*, 1985; Dawes, 1986). This is a distinct advantage over most other carbohydrates which have to be phosphorylated by ATP *in addition* to the energy expended on transport (Muir *et al.*, 1985; Dawes, 1986; Cooper, 1986).

The binding protein sugar transport systems

Some transport activities are abolished by 'cold osmotic shock' (Neu and Heppel, 1965). This process releases periplasmic proteins, whilst conserving the integrity of the inner membrane. Subsequently, 'shock-sensitive' systems were found to involve the essential participation of a periplasmic substrate binding protein in the transport process (reviewed by Anraku, 1978; Silhavy *et al.*, 1978; Furlong, 1986). So far as is known this mechanism is unique to bacteria, and will probably be found in all species exhibiting binding protein-mediated chemotaxis (Ordal, 1985).

Sugar substrates

The sugars that are presently known to enter *E. coli* and *Salmonella typhimurium* by a binding protein mechanism are maltodextrins, maltose, galactose, arabinose, xylose and ribose (*Table 1*). Each system involves different protein components. Certain oligopeptides, amino acids, ions and vitamins utilize analogous transport systems (Furlong, 1986).

Proteins and genes

The periplasmic binding proteins are water-soluble and relatively easy to purify owing to their resistance to heat, pH changes and proteases (Furlong, 1986). The primary structures of the arabinose-, galactose-, ribose- and maltose-binding proteins have been determined by sequencing the pure proteins and/or their genes (see references in Furlong, 1986 and Henderson, 1986). Some have been crystallized, and their three-dimensional structures determined by X-ray diffraction analysis (Gilliland and Quiocho, 1981; Saper and Quiocho, 1983). By this means, a spectacular insight into the precise interactions of substrate -OH residues with the arabinose-binding protein was achieved (Quiocho and Vyas, 1984; Quiocho, 1986). But how could such hydrophilic proteins located in the periplasmic space contribute to the movement of substrate through the hydrophobic inner membrane? Integral membrane proteins, which cooperate with the binding protein, must be required for transport. Evidence for their existence was obtained by the following genetical approaches.

E. coli mutants impaired in the binding protein transport system for each carbohydrate were isolated by a variety of techniques. The location of each mutation on the gene linkage map (Bachmann, 1983) was then determined (*Table 1*). Each sugar transport system mapped in a different place and operated as an independent transcriptional unit, or 'operon'. Complementation analysis revealed the presence in each operon of at least two (arabinose) and up to five (maltose) genes, one coding for the binding protein and at least one (usually two) coding for proteins located in the inner membrane (reviewed by Hengge and Boos, 1983). The critical observation was that mutation of *any one* of the genes profoundly reduced, if not abolished, transport activity (see for example Robbins *et al.*, 1976; Hengge and Boos, 1983). Hence, transport is a cooperative function of a multi-protein complex (*Figure 1*; Hengge and Boos, 1983). A generalization of the roles of the individual protein components can be made, based on the observation that four gene products are usually found.

One protein is cytoplasmic, or loosely associated with the inner surface of the cytoplasmic membrane. Several of these proteins, from different transport systems, have been sequenced *via* their DNA. Despite their different transport substrates, their primary sequences contain extensive homologous regions, some of which correspond to a nucleotide-binding site (Walker *et al.*, 1982; Gilson *et al.*, 1982; Higgins *et al.*, 1985, 1986). These proteins are therefore assumed to be the initial transducers of energy from ATP into the transport process *(Figure 1)*. In the case of the maltose transport system, this protein (MalK), attaches to the inner membrane only when a functional MalG protein is present in the membrane (Hengge and Boos, 1983); MalG is therefore assumed to be the next link in the energization process.

The next two proteins were difficult to identify, purify and characterize, because of their low abundance and hydrophobic nature (Shuman *et al.*, 1980; Hengge and Boos, 1983). However, DNA sequencing of their genes has established the amino acid sequences of five pairs of these proteins (separate transport systems for maltose, ribose, oligopeptides, histidine and phosphate; Higgins *et al.*, 1982; Dassa and Hofnung, 1985; Bell *et al.*, 1986). The primary sequences are not homologous, except perhaps for one short segment about 90 residues from the C-terminal ends (Dassa and Hofnung, 1985), but they are all typical membrane proteins with similar hydropathic profiles (see for example Dassa and Hofnung, 1985). Again, the presence of *both* proteins of each pair is essential for transport, but their precise roles are not yet understood.

The characterized periplasmic binding proteins exhibit similar tertiary structures despite their different substrates − from SO_4^{2-} to disaccharides and oligopeptides (Vyas *et al.*, 1983; Quiocho and Vyas, 1984; Furlong, 1986). A single substrate molecule binds in a deep cleft between two domains joined by a 'hinge' region (Quiocho and Vyas, 1984) and is completely surrounded by the protein, implying that its entry and exit involve a conformational change. The dissociation constants are low, $0.1-5$ μM (Furlong, 1986). A binding protein presumably also 'recognizes' the integral membrane protein(s), chemotactic receptor proteins (Adler, 1975; Ordal, 1985) and, in some cases, an outer membrane component (see below). The knowledge of their detailed three-dimensional structure provides the opportunity for site-directed mutagenesis of individual amino acid residues to elucidate their roles in maintaining protein structure and function (Winter *et al.*, 1982; Zoller and Smith, 1983). In addition, the use of antibodies (Gabay *et al.*, 1985), should soon provide insights into the mechanisms by which the binding proteins interact with the other proteins.

In addition to these components, an outer membrane protein, LamB, assists the binding protein-dependent transport of maltodextrins (Hengge and Boos, 1983). Since it is also the surface receptor for λ phage, now exploited as a vector for genetic engineering, the structure and functions of Lam B proteins have been extensively described elsewhere (reviewed by Hengge and Boos, 1983; Gabay *et al.*, 1985).

The detailed organization of the genes involved in binding protein systems is best known for the maltose and ribose transport systems. A divergent operon, *malB*, is located at 91.5 min on the *E. coli* chromosome with the following gene order (Raibaud *et al.*, 1980):

malG malF malE p/o *malK lamB*

Its expression is positively regulated by the product of the *malT* gene (in the *malA* operon at 74 min), which binds between *malE* and *malK* (Raibaud and Schwartz, 1980). The outer membrane component is coded for by *lamB*, the binding protein by *malE*, the integral membrane components by *malF,G* and the cytoplasmic adenine nucleotide binding component by *malK* (Shuman and Silhavy, 1981; Hengge and Boos, 1983). Since the products of *lamB* and *malE* are much more abundant than those of *malF, malG* and *malK*, there is presumably differential regulation of gene expression, perhaps mediated by intergenic sequences (Clement and Hofnung, 1981; Stern *et al.*, 1984). The DNA sequence of the complete operon is now established (Clement and Hofnung, 1981; Gilson *et al.*, 1982; Duplay *et al.*, 1984; Froshauer and Beckwith, 1984; Dassa and Hofnung, 1985).

The ribose transport system is encoded in an operon at 84 min on the chromosome (Anderson and Cooper, 1970; David and Wiesmeyer, 1970; Iida *et al.*, 1984; Lopilato *et al.*, 1984):

$$p/o \; rbsA \; rbsC \; rbsB \; rbsK$$

The binding protein is coded for by *rbsB*, two presumed membrane components by *rbsC* and *rbsD* and the nucleotide-binding component by *rbsA* (Buckel *et al.*, 1986). The *rbsK* gene encodes ribokinase. Their DNA has recently been sequenced (Bell *et al.*, 1986; Hope *et al.*, 1986).

It is expected that the DNA sequences of more binding protein-dependent sugar transport systems will be determined soon. An interesting feature of the amino acid sequences of the binding proteins for different sugars is that their primary sequences are not strikingly homologous, despite the evidence that their tertiary structures are similar (Quiocho, 1986).

Regulation

The expression of these operons is dependent on the presence of the substrate. In every case the inducer appears to be the chemically unmodified sugar acting from inside the cell (Silhavy *et al.*, 1978). Since cAMP and *crp* protein are necessary for optimal gene expression (Wilson, 1974; Silhavy *et al.*, 1978; Daruwalla *et al.*, 1981; Kolodrubetz and Schleif, 1981; Kosiba and Schleif, 1982; Hengge and Boos, 1983; Hendrickson and Schleif, 1984) these systems are subject to catabolite repression by glucose (see above) or other environmental factors that may influence intracellular cAMP concentrations.

Recently, good evidence was obtained that Enzyme III[glc] inhibited the maltose transport system directly (reviewed by Postma and Lengeler, 1985). This is a mechanism by which glucose, present with maltose in the environment, can directly attenuate maltose entry, and so indirectly impair induction (i.e., 'inducer exclusion'). This inhibitory influence of glucose transport may also be exerted on other binding protein transport systems.

Energetics of binding protein sugar transport systems

The known sugar binding protein transport systems did not mediate H^+ symport and were sensitive to inhibition by arsenate, which depletes the intracellular ATP concentration (Klein and Boyer, 1972; Curtis, 1974; Ferenci *et al.*, 1977; Henderson *et al.*, 1977; Daruwalla *et al.*, 1981; Davis *et al.*, 1984). They were operational when the trans-membrane proton gradient was discharged provided ATP concentrations were maintained by glycolysis (Berger, 1973; Berger and Heppel, 1974; Curtis, 1974).

Additionally, there is indirect and direct evidence for the binding of a nucleotide to one of the protein components (see above; Hobson *et al.*, 1984; Higgins *et al.*, 1985, 1986). The available information therefore strongly suggests that these transport systems are directly energized by ATP without an intervening transmembrane ion gradient. In one case, that of maltose transport, $1-1.2$ mol of ATP are consumed per mol of carbohydrate transported (Muir *et al.*, 1985). It will be important to learn if this stoichiometry is true also for other binding protein transport systems.

Cation-linked sugar transport systems

Mitchell (1963, 1966) realized that a trans-membrane proton gradient generated by respiration could be used to energize nutrient transport as well as ATP synthesis. He proposed that a sugar transport protein could catalyse an obligatory sugar$-H^+$ symport (or the equivalent sugar$-OH^-$ antiport). It would then provide a route for proton re-equilibration across the membrane in competition with ATP synthase, the flagellar motor and other transport systems (Dawes, 1986). In this way the energy released by substrate oxidation and stored in the proton gradient could be transduced to a gradient of [sugar]$_{inside}$ $>>$ [sugar]$_{outside}$ as illustrated in *Figures 1* and *2*.

The following evidence demonstrates the existence of this mechanism in *E. coli* and *S. typhimurium*: the dependence of sugar transport on respiration; its susceptibility to uncoupling agents and ionophores (agents that render the membrane permeable to H^+, K^+ or Na^+); its energization by an artificially generated trans-membrane electrochemical gradient; and the direct observation of sugar-catalysed proton translocation (reviewed by West, 1980; Hengge and Boos, 1983; Kaback, 1986; Henderson and Macpherson, 1986; Henderson, 1986). In anaerobic conditions this mechanism is driven by an electrochemical gradient generated either from ATP (derived from glycolysis), hydrolysed by H^+-ATPase (Dawes, 1986), or by the utilization of alternative electron acceptors (e.g. fumarate and NO_3^- in *E. coli*; Ingledew and Poole, 1984), or possibly by a chemical gradient of protons generated from lactate efflux (Michels *et al.*, 1979; ten Brink and Konings, 1982).

Sugar substrates

The sugars that are presently known to enter *E. coli* by a cation-linked mechanism are lactose (H^+), melibiose (Na^+), galactose (H^+), arabinose (H^+), xylose (H^+), L-fucose (H^+) and L-rhamnose (H^+) (West, 1970; Kaback, 1986; Wilson *et al.*, 1986; Henderson and Macpherson, 1986). How widely this mechanism for sugar transport is disseminated amongst bacterial species is not known, but our own measurements on the enterobacteria indicate a somewhat sporadic occurrence; the lactose$-H^+$ and xylose$-H^+$ systems are absent from *S. typhimurium*, and only the lactose$-H^+$ and rhamnose$-H^+$ systems are found in *Erwinia carotovora* (unpublished data). However, sugar$-$cation transport systems do occur in the eukaryotes, including yeasts (Eddy, 1982), algae (Komor, 1973), higher plants (Komor, 1977; Hutchings, 1978) and, of course, mammals (Schultz and Curran, 1970).

Each of the systems transports related sugar analogues as well as the physiological substrate. The substrates for each are summarized in *Table 1*.

Proteins and genes

In contrast to the other transport systems, proton symporters have only one protein component. The most extensively studied is the lactose transporter of *E. coli*, LacY (Kaback, 1986). LacY has been purified to apparent homogeneity (Newman *et al.*, 1981) and its amino acid sequence established by sequencing the DNA of the *lacY* gene (Büchel *et al.*, 1980). Some progress has been made in establishing the topology of the protein and identifying some of the residues involved in the transport process (Brooker and Wilson, 1985; Markgraf *et al.*, 1985; Sarkar *et al.*, 1986).

Particularly sensitive labelling techniques were required to identify each of the proteins responsible for H^+−arabinose, H^+−galactose, H^+−xylose and Na^+−melibiose transport (reviewed by Henderson and Macpherson, 1986). They had apparent molecular weight values of about 36 000, 37 000, 39 000 and 30 000, respectively as determined by SDS−polyacrylamide gel electrophoresis (Macpherson *et al.*, 1982, 1984; E.O.Davis and P.J.F.Henderson, unpublished data; Hanatani *et al.*, 1983). However, sequencing of the corresponding genes revealed true molecular weight values in the range 51 000−54 000 (Yazyu *et al.*, 1984; E.O.Davis and M.C.J.Maiden, unpublished). This discrepancy appears to be common for integral membrane proteins.

The genes coding for each sugar−cation transporter are scattered around the *E. coli* chromosome (*Table 1*). Those for LacY and MelB occur in operons (at 8 min and 92 min, respectively) that also contain the genes encoding the enzyme(s) required for entry of the sugar into metabolism (*Figure 3*; Bachmann, 1983). However, each of the genes for galactose−H^+ (*galP*, 64 min), arabinose−H^+ (*araE*, 61 min) and xylose−H^+ (*xylE*, 91 min) are in operons separated from those encoding the genes for sugar metabolism and the other binding protein transport system (*Figure 3*; Henderson and Macpherson, 1986). This separation can simplify the mutagenesis and/or cloning of the individual transport genes (Jones-Mortimer and Henderson, 1986). Preliminary evidence indicates that the rhamnose−H^+ transport gene maps near the

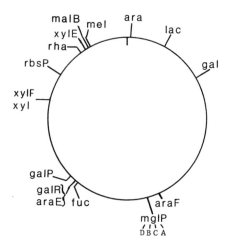

Figure 3. The location of *E. coli* genes involved in sugar transport and metabolism. The nomenclature of the genes is explained in the text and by Bachmann (1983). The mark on the inside of the circular linkage map is 0 min, rising to 100 min in a clockwise direction.

rha operon at 87 min (J.A.R.Muiry, unpublished data), but the fucose−H^+ transport gene has not yet been located (the *fuc* operon is at 60 min; Bachmann, 1983).

One example of experimental progress from mutation of the *araE* gene to the elucidation of its product's protein structure is given further below.

Regulation

The expression of an operon for any one of the cation-linked transport systems is dependent on the presence of the appropriate substrate. The inducer is often chemically unmodified sugar acting from inside the cell, except for LacY where it is internal allo-lactose (Jobe and Bourgeois, 1972) and for the L-fucose operon, where it is probably L-fuculose-1-phosphate (Bartkus and Mortlock, 1986). Our knowledge of the regulatory mechanisms varies greatly. The regulation of the *lac* operon is, of course, extremely well characterized as a 'negatively' regulated system (see for example Glass, 1984; Silhavy *et al.*, 1984). Recently, the 'positive' (and negative?) regulation of the arabinose transport genes, *araE* and *araF*, by the *araC* gene product has been elegantly investigated by Schleif and co-workers (Kolodrubetz and Schleif, 1981; Kosiba and Schleif, 1982; Stoner and Schleif, 1983; Hendrickson and Schleif, 1984). Both are subject to catabolite repression, but *araE* rather less so than *araF* (Daruwalla *et al.*, 1981). It remains an interesting question whether expression of *araE* and *araF* is regulated in a reciprocal way − can one transport system sense the level of expression of the other? A suggestion of such a 'handshake' phenomenon was made for the two galactose transport systems by Wilson (1974)

The promoter regions of the *melB* gene, the *xylE* gene and, putatively, the *galP* gene have recently been sequenced (Yazyu *et al.*, 1983; Davis and Henderson, in preparation; D.C.M.Moore, unpublished), but an understanding of their detailed regulatory features awaits experiments that measure their binding of the regulatory proteins, including *crp* (see for example Hendrickson and Schleif, 1984). Since strategies are well established for investigating regulation of prokaryote gene expression, it is likely that much more information will accrue in the near future.

A notable discovery is that unphosphorylated Enzyme III^{glc} directly inhibits activity of the LacY transport protein (reviewed by Saier, 1985). Once again glucose transport is able to attenuate the activity and exclude the inducer of a rival transport system. It seems likely that glucose transport directly influences the activity of other cation-linked transport systems (Okada *et al.*, 1981).

Energetics of cation-linked transport systems

Proton-linked transport systems are characterized by the dependence of transport into vesicles on respiration and its sensitivity to uncoupling agents and ionophores (see above). In whole cells the transport activity is relatively insensitive to arsenate (i.e. intracellular ATP concentrations) and a sugar-promoted pH change can be observed when substrate is added to energy-depleted cells (West, 1970; Henderson and Macpherson, 1986). Na^+-linked transport systems display similar properties, and an additional dependence on added Na^+ ions.

Owing to the availability of cation-sensitive electrodes it is easy to measure, continuously and quantitatively, cation movements that accompany sugar transport into bacterial cells (Wilson *et al.*, 1986; Henderson and Macpherson, 1986). It is much more difficult to measure accurately the sugar movement, since experiments must be done with high concentrations under de-energized conditions using radioactive sugars assayed by discontinuous sampling. Nevertheless, our measurements of pH changes elicited by arabinose, xylose, galactose, fucose and rhamnose indicate an H^+/sugar stoichiometry of $0.6-1.0$, similar to values of 1.0, or up to 2.0, reported for the best-studied lactose transporter (West and Mitchell, 1973; Ahmed and Booth, 1983; reviewed by Hengge and Boos, 1983). Muir *et al.* (1985) pointed out that this would imply a transport 'cost' of $0.3-0.6$ ATP, given that $3H^+$ are required for the synthesis of ATP from ADP in *E. coli* (Kashket, 1982; Vink *et al.*, 1984). They obtained evidence that about 0.5 ATP was indeed consumed by transport of lactose by LacY, compared with $1-1.2$ ATP consumed per maltose transported by MalB (the binding protein transport system discussed above). This is the best evidence available that cation-linked transport systems are energetically cheaper than binding protein systems.

Transport of melibiose by the MelB system can apparently use H^+ ions if Na^+ ions are absent (Tsuchiya and Wilson, 1978; Tsuchiya *et al.*, 1980; Damiano-Forano *et al.*, 1986). It is not known whether this adaptability confers any selective advantage or affects growth parameters.

Characterization of the arabinose–H^+ transporter of *E. coli*

This section illustrates the strategy outlined above for the investigation of membrane proteins, with reference to our recent work on L-arabinose–H^+ symport in *E. coli*. Arabinose–H^+ symport was shown to be catalysed by a transport system corresponding to the *araE* gene (Daruwalla *et al.*, 1981). The product of *araE* was a single integral membrane protein of apparent molecular weight 36 000 (Macpherson *et al.*, 1981).

Mutagenesis and cloning of the araE gene

The phage λp*lac*Mu1 (Silhavy *et al.*, 1984; Bremer *et al.*, 1984) was used to mutate *araE*. This phage inserts randomly into the *E. coli* chromosome and, depending on the position and orientation of the insertion, confers a Lac⁺ phenotype. The Lac⁺ phenotype is under the control of the promoter of the target gene, and a hybrid protein of the N-terminal of the target gene and β-galactosidase is produced. Lac fusions are particularly useful in the study of proton symport as they confer a lactose–H^+ symport activity, which is a valuable control. It is simple to isolate phage and adjacent host DNA from strains mutagenized by λp*lac*Mu1.

The strain with the insert into *araE*, JM2443, was characterized and its genetic structure confirmed by the following observations.

(i)   The fusion was co-transducible with *lysA*, demonstrating the location of the insertion at 61 min on the linkage map.

(ii)   The Lac⁺ phenotype and the fusion protein were arabinose inducible.

(iii)   The strain had lost arabinose–H^+ symport activity and gained lactose–H^+ symport activity.

(iv)   The β-galactosidase activity was associated with the cell membrane indicating that the insertion was into a gene coding for a membrane protein.

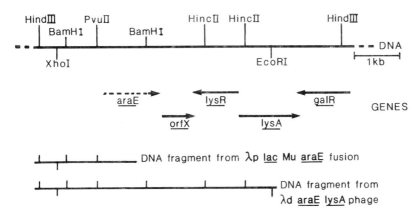

Figure 4. The 61 min region of the *E. coli* chromosome. A restriction map of the region is shown with the relative positions of the genes marked (Stragier and Patte, 1983). The approximate location of *araE* is shown, within the subsequently-cloned *Xho*I−*Eco*RI fragment (*Figure 5*).

Phage carrying the fusion were obtained by u.v. irradiation of the lysogen and screening for phage with an arabinose-inducible β-galactosidase activity. These phage carried the *araE* promoter and coded for the fusion protein. A restriction enzyme site map of this phage showed that it carried 8−10 kb of host DNA. Comparing this map with similar maps of the *E. coli* chromosome and the phage λd(*araE*$^+$*lysA*$^+$*galR*$^+$) (Macpherson *et al.*, 1981) established the location and direction of transcription of *araE* (*Figure 4*).

A 4.7-kb *Xho*I−*Eco*RI restriction fragment containing *araE* was subcloned from the phage λd(*araE*$^+$*lysA*$^+$*galR*$^+$) into the multicopy plasmid vector pBR322 (*Figure 5*). This enabled the production of large quantities of *araE* DNA. From this plasmid (pMM25) a smaller piece of 2.8 kb was isolated and used for subcloning into sequencing and expression vectors (*Figure 5*, see below).

Overexpression of AraE

A simple way of increasing the level of expression of a protein is to increase the concentration of mRNA coding for the gene product. This can be achieved by using a promoter which initiates the transcription more readily than the wild-type promoter. The phage λ promoter p_L is one of the most efficient known (Rosenberg *et al.*, 1982). It has the additional advantage of being under tight negative control by the phage *cI* gene; membrane proteins can be toxic to the cell when over-produced, hence an efficient off−on 'switch' is essential. The p_L promoter can be 'switched on' either by heat shock of a host with *cI*857 containing the plasmid, or by the addition of nalidixic acid to a *cI*$^+$ strain (Mott *et al.*, 1985). Using the recombinant plasmid pMM27 (*Figure 5*), which contained the 2.8 kb restriction fragment in the correct orientation for the expression of *araE* from P_L, we have raised the level of AraE to 10−15% of the inner membrane proteins, an amplification of about 20-fold.

'Batch type' culture conditions are the best way of over-producing this protein because it is toxic; other non-toxic products could be produced continuously (and with a much higher yield) under chemostat culture conditions.

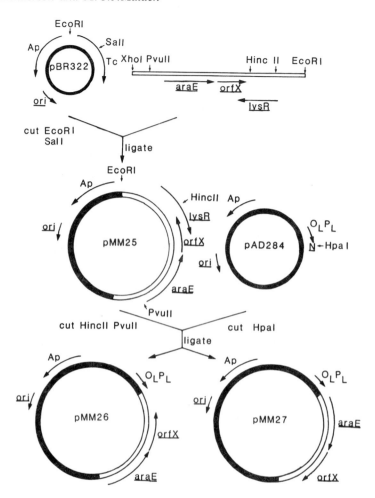

Figure 5. Strategy for cloning and amplifying expression of the *araE* gene. A *Xho*I−*Eco*RI fragment of 4.8 kb was sub-cloned from λd (*galR⁺lysA⁺araE⁺*) phage (Macpherson *et al.*, 1981) into the *Eco*RI and *Sal*I sites of pBR322. This plasmid (pMM 25) contained *araE*, *orfX*, *lysR* and part of *lysA*. *E. coli* strain MM23 (*araE araF*) acquired arabinose−H⁺ symport activity when transformed with pMM25, demonstrating the presence of an intact *araE* gene on the latter. A smaller *Hinc*II−*Pvu*II fragment of 2.8 kb also carried H⁺−arabinose symport activity. This was used to generate sequence data by the method of Sanger *et al.* (1980), and was cloned into a λP$_L$ expression vector (pAD284, from Asis Das). In plasmid pMM27 *araE* was orientated in such a way as to allow its expression from the λP$_L$; in plasmid pMM26 *araE* was in the opposite orientation.

DNA sequencing of araE

The 2.8-kb fragment containing *araE* was circularized, sonicated and size fractionated to generate a random population of pieces of DNA 300−600 bases in size. These were cloned into the M13 sequencing vectors (Messing *et al.*, 1981) and sequenced by the method of Sanger *et al.* (1981). A contiguous sequence of the 2.8-kb fragment was assembled. The open reading frame corresponding to *araE* was found and translated to give the amino acid sequence of AraE (*Figure 6*).

THE ARABINOSE-H⁺ SYMPORT PROTEIN

```
                5                   10                  15                  20
Met Val Thr Ile Asn Thr Glu Ser Ala Leu Thr Pro Arg Ser Leu Arg Asp Thr Arg Arg
                25                  30                  35                  40
Met Asn Met Phe Val Ser Val Ala Ala Ala Val Ala Gly Leu Leu Phe Gly Leu Asp Ile
                45                  50                  55                  60
Gly Val Ile Ala Gly Ala Leu Pro Phe Ile Thr Asp His Phe Val Leu Thr Ser Arg Leu
                65                  70                  75                  80
Gln Glu Trp Val Val Ser Ser Met Met Leu Gly Ala Ala Ile Gly Ala Leu Phe Asn Gly
                85                  90                  95                 100
Trp Leu Ser Phe Arg Leu Gly Arg Lys Tyr Ser Leu Met Ala Gly Ala Ile Leu Phe Val
                105                 110                 115                120
Leu Gly Ser Ile Gly Ser Ala Phe Ala Thr Ser Val Glu Met Leu Ile Ala Ala Arg Val
                125                 130                 135                140
Val Leu Gly Ile Ala Val Gly Ile Ala Ser Tyr Thr Ala Pro Leu Tyr Leu Ser Glu Met
                145                 150                 155                160
Ala Ser Glu Asn Val Arg Gly Lys Met Ile Ser Met Tyr Gln Leu Met Val Thr Leu Gly
                165                 170                 175                180
Ile Val Leu Ala Phe Leu Ser Asp Thr Ala Phe Ser Tyr Ser Gly Asn Trp Arg Ala Met
                185                 190                 195                200
Leu Gly Val Leu Ala Leu Pro Ala Val Leu Leu Ile Ile Leu Val Val Phe Leu Pro Asn
                205                 210                 215                220
Ser Pro Arg Trp Leu Ala Glu Lys Gly Arg His Ile Glu Ala Glu Glu Val Leu Arg Met
                225                 230                 235                240
Leu Arg Asp Thr Ser Glu Lys Ala Arg Glu Glu Leu Asn Glu Ile Arg Glu Ser Leu Lys
                245                 250                 255                260
Leu Lys Gln Gly Gly Trp Ala Leu Phe Lys Ile Asn Arg Asn Val Arg Arg Ala Val Phe
                265                 270                 275                280
Leu Gly Met Leu Leu Gln Ala Met Gln Gln Phe Thr Gly Met Asn Ile Ile Met Tyr Tyr
                285                 290                 295                300
Ala Pro Arg Ile Phe Lys Met Ala Gly Phe Thr Thr Thr Glu Gln Gln Met Ile Ala Thr
                305                 310                 315                320
Leu Val Val Gly Leu Thr Phe Met Phe Ala Thr Phe Ile Ala Val Phe Thr Val Asp Lys
                325                 330                 335                340
Ala Gly Arg Lys Pro Ala Leu Lys Ile Gly Phe Ser Val Met Ala Leu Gly Thr Leu Val
                345                 350                 355                360
Leu Gly Tyr Cys Leu Met Gln Phe Asp Asn Gly Thr Ala Ser Ser Gly Leu Ser Trp Leu
                365                 370                 375                380
Ser Val Gly Met Thr Met Met Cys Ile Ala Gly Tyr Ala Met Ser Ala Ala Pro Val Val
                385                 390                 395                400
Trp Ile Leu Cys Ser Glu Ile Gln Pro Leu Lys Cys Arg Asp Phe Gly Ile Thr Cys Ser
                405                 410                 415                420
Thr Thr Thr Asn Trp Val Ser Asn Met Ile Ile Gly Ala Thr Phe Leu Thr Leu Leu Asp
                425                 430                 435                440
Ser Ile Gly Ala Ala Gly Thr Phe Trp Leu Tyr Thr Ala Leu Asn Ile Ala Phe Val Gly
                445                 450                 455                460
Ile Thr Phe Trp Leu Ile Pro Glu Thr Lys Asn Val Thr Leu Glu His Ile Glu Arg Lys
                465                 470
Leu Met Ala Gly Glu Lys Leu Arg Asn Ile Gly Val
```

Figure 6. The amino acid sequence of the arabinose–H^+ symport protein. The correct reading frame and predicted amino acid sequence were established within the DNA sequence of the *Hinc*II–*Pvu*II fragment (Figure 5).

The AraE protein sequence

AraE is a highly hydrophobic (66% unpolar residues) protein with a calculated molecular weight of 51 683. The protein consists of alternating hydrophobic and hydrophilic regions, a characteristic of many membrane proteins (Kyte and Doolittle, 1982). This suggests a protein structure similar to that of bacteriorhodopsin (Henderson and Unwin, 1975) with membrane-spanning loops passing through the membrane, from side to side. The sequence can be used to propose a tentative model of the protein's structure, which will be of use in the design of experiments to elucidate its actual three-dimensional arrangement.

Homology of AraE with other *E. coli* transport proteins

In our laboratory the *xylE* gene coding for xylose−H$^+$ symport has been mapped, cloned and sequenced (Davis *et al.*, 1984; Davis and Henderson, in preparation). XylE is homologous with AraE (Maiden *et al.*, 1987). By comparing the sequences of two proteins of known similar function it is possible to identify residues of importance in the transport process. *In vitro* and *in vivo* mutagenesis will then establish whether concepts of the molecular mechanism developed from such comparisons are valid. AraA and XylE are also homologous with the *E. coli* citrate transporter (Ishiguro and Sato, 1985; Sasatsu *et al.*, 1985), and there are compelling similarities between these three proteins and a tetracycline resistance factor which exports tetracycline from the *E. coli* cell (McMurray *et al.*, 1980; Hillen and Schallmeier, 1983). Three regions in which patterns of residues are conserved have been identified, despite the diversity of transport substrate. These presumably form critical structural features of the transport proteins.

Homology of AraE with a glucose transporter from man

Mueckler *et al.* (1985) obtained, by sequencing a cDNA clone, the primary structure of a glucose transporter from a human hepatoma cell line. This protein was highly homologous (possibly identical) with the erythrocyte glucose transporter. Unexpectedly, this protein is as homologous to AraE and XylE as they are to each other (Maiden *et al.*, 1987). The homologies are such that structural data from each will be applicable to the others. Thus, the determination of the amino acid sequences of these proteins via DNA sequencing has unified two quite disparate fields of study, enabling techniques and information from each to be applied to the other.

It is interesting to speculate on the evolutionary origin of this homology. Was there an ancient precursor of these proteins in existence before the divergence of prokaryotes and eukaryotes? Or has there been a more recent exchange of genetic information between the mammal and its gut symbiont? It seems unlikely that such similar proteins could arise through convergent evolution, especially when *E. coli* contains other sugar−proton transport proteins (LacY and MelB; see above) that are not apparently homologous with AraE and XylE (or the glucose carrier).

The use of recombinant DNA technology has enabled us to gain insights into transport processes impossible by other techniques, and has challenged us with many exciting questions to guide future research that will be of value to biotechnology.

Acknowledgements

Our laboratory research is supported by SERC grant GR/C/34977 and by equipment grants from the Smith-Kline Foundation and the Wellcome Trust. M.C.J.M. thanks SERC and Sidney Sussex College, Cambridge, for studentships and MRC for a fellowship.

References

Adler,J. (1975) Chemotaxis in bacteria. *Annu. Rev. Biochem.*, **44**, 341−356.

Ahmed,S. and Booth,I.C. (1983) The effect of β-galactosides on the protonmotive force and growth of *Escherichia coli*. *J. Gen. Microbiol.*, **129**, 2521−2529.

Anderson,A. and Cooper,R.A. (1970) Biochemical and genetical studies on ribose catabolism in *Escherichia coli* K12. *J. Gen. Microbiol.*, **62**, 335−339.

Anraku,Y. (1978) Active transport of amino acids. In *Bacterial Transport*. Rosen,B.P. (ed.), Marcel Dekker Inc., New York, pp. 171−219.

Bachmann,B.J. (1983) Linkage map of *Escherichia coli* K12 (Edition 7). *Microbiol. Rev.*, **47**, 180−230.

Bartkus,J.M. and Mortlock,R.P. (1986) Isolation of a mutation resulting in constitutive synthesis of L-fucose catabolic enzymes. *J. Bacteriol.*, **165**, 710−714.

Bell,A.W., Buckel,S.D., Groarke,J.M., Hope,J.N., Kingsley,D.H. and Hermodson,M.A. (1986) Nucleotide sequences of the *rbsD, rbsA*, and *rbsC* genes of *Escherichia coli* K12. *J. Biol. Chem.*, **261**, 7652−7658.

Berger,E.A. (1973) Different mechanisms of energy coupling for the active transport of proline and glutamine in *Escherichia coli*. *Proc. Natl. Acad. Sci. USA*, **70**, 1510−1518.

Berger,E.A. and Heppel,L.A. (1974) Different mechanisms of energy coupling for the shock sensitive and shock resistant amino acid permeases of *Escherichia coli*. *J. Biol. Chem.*, **249**, 7747−7755.

de Boer,M., Brockhuizen,C.P. and Postma,P.W. (1986) Regulation of glycerol kinase by EnzymeIII^{Glc} of the PEP:carbohydrate phosphotransferase system. *J. Bacteriol.*, **167**, 393−395.

Bramley,H.A. and Kornberg,H.L. (1987) Sequence of the *bgl* gene coding for EnzymeII of the β-glucoside transport system of *Escherichia coli*. *J. Gen. Microbiol.*, **133**, 563−573.

Bremer,E., Silhavy,T.J., Weisemann,J.M. and Weinstock,G.M. (1984) λ*plac*Mu: a transposable derivative of bacteriophage for creating lacZ protein fusions in a single step. *J. Bacteriol.*, **158**, 1084−1093.

ten Brink,B. and Konings,W.N. (1982) The electrochemical proton gradient and lactate concentration gradient in *Streptococcus cremoris* grown in batch culture. *J. Bacteriol.*, **152**, 682−686.

Brooker,R.J. and Wilson,T.H. (1985) Isolation and nucleotide sequencing of lactose carrier mutants that transport maltose. *Proc. Natl. Acad. Sci. USA*, **82**, 3959−3963.

Büchel,D.E., Gronenborn,B. and Müller-Hill,B. (1980) Sequence of the lactose permease gene. *Nature*, **283**, 541−545.

Buckel,W. (1986) Biotin-dependent decarboxylases as bacterial sodium pumps. Purification and reconstitution of glutaconyl-CoA decarboxylase from *Acidaminococcus fermentans*. *Methods Enzymol.*, **125**, 547−558.

Buckel,S.D., Bell,A.W., Mohana Roao,J.K. and Henderson,M.A. (1986) An analysis of the structure of the product of the *rbsA* gene of *Escherichia coli* K12. *J. Biol. Chem.*, **261**, 7659−7662.

Button,D.K. (1985) Kinetics of nutrient-limited transport and microbial growth. *Microbiol. Rev.*, **49**, 270−297.

Cairney,J., Booth,I.R. and Higgins,C.F. (1986) Osmoregulation of gene expression in *Salmonella typhimurium*: *proU* encodes an osmotically induced betaine transport system. *J. Bacteriol.*, **164**, 1224−1232.

Clément,J.M. and Hofnung,M. (1981) Gene sequences of the λ receptor, an outer membrane protein of *Escherichia coli* K12. *Cell*, **27**, 507−514.

Cooper,R.A. (1986) Convergent pathways of sugar catabolism in bacteria. In *Carbohydrate Metabolism in Cultured Cells*. Morgan,M.J. (ed.), Plenum Press, London, pp. 461−491.

Curtis,S.J. (1974) Mechanism of energy coupling for transport of D-ribose in *Escherichia coli*. *J. Bacteriol.*, **120**, 295−303.

Damiano-Forano,E., Bassilana,M. and Leblanc,G. (1986) Sugar binding properties of the melibiose permease in *Escherichia coli* membrane vesicles. *J. Biol. Chem.*, **261**, 6893−6899.

Daruwalla,K.R., Paxton,A.T. and Henderson,P.J.F. (1981) Energisation of the transport systems for arabinose and comparison with galactose transport in *Escherichia coli*. *Biochem. J.*, **200**, 611−627.

Dassa,E. and Hofnung,M. (1985) Sequence of the gene *malG* in *E. coli* K12. *EMBO J.*, **4**, 2287−2293.

David,J.D. and Wiesmeyer,H. (1970) Regulation of ribose catabolism in *Escherichia coli*: the ribose catabolic pathway. *Biochim. Biophys. Acta*, **208**, 45−55.

Davis,E.O. (1985) Xylose transport in *Escherichia coli*. Ph.D. Thesis, University of Cambridge.

Davis,E.O., Jones-Mortimer,M.C. and Henderson,P.J.F. (1984) Location of a structural gene for xylose-H$^+$ symport at 91 min on the linkage map of *Escherichia coli* K12. *J. Biol. Chem.*, **259**, 1520−1525.

Dawes,E.A. (1986) *Microbial Energetics*. Blackie, London.

di Renzio,J.M., Nakamura,K. and Inouye,M. (1978) The outer membrane proteins of gram-negative bacteria: biosynthesis, assembly and functions. *Annu. Rev. Biochem.*, **47**, 481−532.

Dimroth,P. (1986) Preparation, characterisation and reconstitution of oxaloacetate decarboxylase from *Klebsiella aerogenes*, a sodium pump. *Methods Enzymol.*, **125**, 530–540.

Duplay,P., Bedouelle,H., Fowler,A., Zabin,I., Sairin,W. and Hofnung,M. (1984) Sequences of the *malE* gene and of its product the maltose binding protein of *Escherichia coli* K12. *J. Biol. Chem.*, **259**, 10606–10613.

Eddy,A.A. (1982) Mechanisms of solute transport in selected microorganisms. *Adv. Microb. Physiol.*, **23**, 1–78.

Engel,A., Massalski,A., Schindler,H., Dorset,D.L. and Rosenbusch,J.P. (1985) Porin channel triplets merge into single outlets in *Escherichia coli* outer membranes. *Nature*, **317**, 643–645.

Epstein,W. (1982) The Kdp system: a bacterial K^+ transport ATPase. *Curr. Top. Membr. Transp.*, **23**, 153–175.

Ferenci,T., Boos,W., Schwartz,M. and Scmeclman,S. (1977) Energy coupling of the transport system of *Escherichia coli* dependent on maltose binding proteins. *Eur. J. Biochem.*, **75**, 187–195.

Froshauer,S. and Beckwith,J. (1984) The nucleotide sequence of the gene for *malF* protein, an inner membrane component of the maltose transport system of *Escherichia coli*. *J. Biol. Chem.*, **259**, 10896–10903.

Furlong,C.E. (1986) Binding protein-dependent active transport in *Escherichia coli* and *Salmonella typhimurium*. *Methods Enzymol.*, **125**, 279–289.

Futai,M. and Kanazawa,H. (1983) Structure and function of proton-translocating adenosine triphosphatase (F_0F_1). *Microbiol. Rev.*, **47**, 285–312.

Gabay,J., Schenkman,S., Desaymond,C. and Schwartz,M. (1985) Monoclonal antibodies and the structure of bacterial membrane proteins. In *Monoclonal Antibodies Against Bacteria*. Macario,A.S.L. and de Macario,C. (eds), Academic Press, New York, pp. 249–282.

Glass,R.E. (1982) Gene function in *E. coli* and its heritable elements. Croom Helm, London.

Gilson,E., Higgins,C.F., Hofnung,M., Ames,G.F.-L. and Nikaido,J. (1982) Extensive homology between membrane-associated components of histidine and maltose transport systems of *Salmonella typhimurium* and *Escherichia coli*. *J. Biol. Chem.*, **257**, 9915–9918.

Gilliland,G.L. and Quiocho,F.A. (1981) Structure of the L-arabinose-binding protein from *E. coli* at 2.4 Å resolution. *J. Mol. Biol.*, **146**, 341–362.

Hanatani,M., Yazyu,H., Shiota-Niiya,S., Moriyama,Y., Kanazawa,H., Futai,M. and Tsuchiya,T. (1984) Physical and genetic characterisation of the melibiose operon and identification of the gene products in *Escherichia coli*. *J. Biol. Chem.*, **259**, 1807–1812.

Heller,K.B., Lin,E.C.C. and Wilson,T.H. (1980) Substrate specificity and transport properties of the glycerol facilitator of *Escherichia coli*. *J. Bacteriol.*, **144**, 274–278.

Henderson,P.J.F. (1980) The inter-relationship between proton-coupled and binding protein dependent transport systems in bacteria. *Biochem. Soc. Trans.*, **8**, 678–679.

Henderson,P.J.F. (1985) Statistical analysis of enzyme kinetic data. In *Techniques in Protein and Enzyme Biochemistry*. Elsevier, Ireland, Vol. B1/11, Supplement BS114.

Henderson,P.J.F. (1986) Active transport of sugars into *Escherichia coli*. In *Carbohydrate Metabolism in Cultured Cells*. Morgan,M.J. (ed.), Plenum Press, London, pp. 409–460.

Henderson,P.J.F. and Macpherson,A.J.S. (1986) Assay, genetics, proteins, and reconstitution of proton-linked galactose, arabinose and xylose transport systems of *Escherichia coli*. *Methods Enzymol.*, **125**, 387–429.

Henderson,R. and Unwin,P.N.T. (1975) Three-dimensional model of purple membrane obtained by electron microscopy. *Nature New Biol.*, **257**, 28–32.

Henderson,P.J.F., Giddens,R.A. and Jones-Mortimer,M.C. (1977) Transport of galactose, glucose and their molecular analogues by *Escherichia coli* K12. *Biochem. J.*, **162**, 309–320.

Hendrickson,W. and Schleif,R. (1984) Regulation of the *Escherichia coli* L-arabinose operon studied by gel electrophoresis DNA binding assay. *J. Mol. Biol.*, **174**, 611–628.

Hengge,R. and Boos,W. (1983) Maltose and lactose transport in *Escherichia coli*, examples of two different types of concentrative transport systems. *Biochim. Biophys. Acta*, **737**, 443–478.

Higgins,C.F., Haag,P.D., Nikaido,H., Ardeshir,F., Garcia,G. and Ames,G.F.-L. (1982) Complete nucleotide sequence of the histidine transport operon of *Salmonella typhimurium*. *Nature*, **298**, 723–727.

Higgins,C.F., Miles,I.D., Whalley,K. and Jamieson,D.J. (1985) Nucleotide binding by membrane components of bacterial periplasmic binding protein dependent transport systems. *EMBO J.*, **4**, 1033–1040.

Higgins,C.F., Hiles,I.D., Salmond,G.P.C., Gill,D.R., Downie,J.A., Evans,I.J., Holland,I.B., Gray,L., Buckel,S.D., Bell,A.W. and Hermodson,M.A. (1986) A family of related ATP-binding subunits coupled to many distinct biological processes in bacteria. *Nature*, **323**, 448–450.

Hillen,W. and Schallmeier,K. (1983) Nucleotide sequence of the Tn10 encoded tetracycline resistance gene. *Nucleic Acids Res.*, **11**, 525–539.

Hobson,A.C., Weatherwax,R. and Ames,G.F.-L. (1984) ATP binding sites in the membrane components of histidine permease, a periplasmic transport system. *Proc. Natl. Acad. Sci. USA*, **81**, 7333−7337.

Hope,J.N., Bell,A.W., Hermodson,M.A. and Groarke,J.M. (1986) Ribokinase from *Escherichia coli* K12. *J. Biol. Chem.*, **261**, 7663−7668.

Hutchings,V.M. (1978) Sucrose and proton cotransport in *Ricinus* cotyledons. *Planta*, **138**, 229−235.

Iida,A., Harayama,S., Iino,T. and Hazelbauer,G.L. (1984) Molecular cloning and characterisation of genes required for ribose transport and utilisation in *Escherichia coli* K12. *J. Bacteriol.*, **158**, 674−682.

Ingledew,W.S. and Poole,R.K. (1984) The respiratory chains of *Escherichia coli*. *Microbiol. Rev.*, **48**, 222−271.

Ishiguro,N. and Sato,G. (1985) Nucleotide sequence of the gene determining plasmid-mediated citrate utilisation. *J. Bacteriol.*, **164**, 977−982.

Jones-Mortimer,M.C. and Henderson,P.J.F. (1986) Use of transposons to isolate and characterise mutants lacking membrane proteins, illustrated by the sugar transport systems of *Escherichia coli*. *Methods Enzymol.*, **125**, 157−180.

Jobe,A. and Bourgeois,S. (1972) Lac repressor operator interaction. VI. The natural inducer of the *lac* operon. *J. Mol. Biol.*, **69**, 397−408.

Kaback,H.R. (1986) Proton electrochemical gradients and active transport: the saga of *lac* permease. *Ann. N.Y. Acad. Sci.*, **456**, 291−304.

Kagawa,Y. (1978) Reconstitution of the energy transformer, gate and channel, subunit reassembly, crystalline ATPase and ATP synthesis. *Biochim. Biophys. Acta*, **505**, 45−93.

Kashket,E.R. (1982) Stoichiometry of the H^+-ATPase of growing and resting, aerobic *Escherichia coli*. *Biochemistry*, **21**, 5534−5538.

Kleffel,B., Garavito,R.M., Baumeister,W. and Rosenbusch,J.P. (1985) Secondary structure of a channel-forming protein: porin from *E. coli* outer membranes. *EMBO J.*, **4**, 1589−1592.

Klein,W.L. and Boyer,P.D. (1972) Energisation of active transport by *Escherichia coli*. *J. Biol. Chem.*, **247**, 7257−7265.

Koch,A.L. (1971) The adaptive responses of *Escherichia coli* to a feast and famine existence. *Adv. Microb. Physiol.*, **6**, 147−217.

Kolodrubetz,D. and Schleif,R. (1981) Regulation of L-arabinose transport operons in *Escherichia coli*. *J. Mol. Biol.*, **151**, 215−227.

Komor,E. (1973) Proton-coupled hexose transport in *Chlorella vulgaris*. *FEBS Lett.*, **38**, 16−18.

Komor,E. (1977) Sucrose uptake by cotyledons of *Ricinus communial*: characteristics, mechanism, regulation. *Planta (Berl.)*, **137**, 119−131.

Kornberg,H.L. (1976) Genetics in the study of carbohydrate transport in bacteria. *J. Gen. Microbiol.*, **96**, 1−16.

Kornberg,H.L. (1986) The roles of HPr and FPr in the utilisation of fructose by *Escherichia coli*. *FEBS Lett.*, **194**, 12−15.

Kosiba,B.E. and Schleif,R. (1982) Arabinose inducible promoter from *Escherichia coli*. *J. Mol. Biol.*, **156**, 53−56.

Kundig,W. (1974) Molecular interactions in the bacterial phosphoenol pyruvate-phosphotransferase system (PTS). *J. Supramol. Struct.*, **2**, 695−714.

Kyte,J. and Doolittle,R.F. (1982) A simple method for displaying the hydropathic character of a protein. *J. Mol. Biol.*, **157**, 105−132.

Lee,C.A. and Saier,M.H. (1983) Mannitol-specific Enzyme II of the bacterial phosphotransferase system. *J. Biol. Chem.*, **258**, 10761−10767.

Lin,E.C.C. (1986) Glycerol facilitator in *Escherichia coli*. *Methods Enzymol.*, **125**, 467−473.

Lopilato,J.E., Garwin,J.L., Emr,S.D., Silhavy,T.J. and Beckwith,J.R. (1984) D-Ribose metabolism in *Escherichia coli* K12: genetics, regulation and transport. *J. Bacteriol.*, **158**, 665−673.

Macpherson,A.J.S., Jones-Mortimer,M.C. and Henderson,P.J.F. (1981) Identification of the AraE transport protein of *Escherichia coli*. *Biochem. J.*, **196**, 269−283

Macpherson,A.J.S., Jones-Mortimer,M.C., Home,P. and Henderson,P.J.F. (1984) Identification of the GalP galactose transport protein of *Escherichia coli*. *J. Biol. Chem.*, **258**, 4390−4396.

Maiden,M.C.J., Davis,E.O., Baldwin,S.A., Moore,D.C.M. and Henderson,P.J.F. (1987) Mammalian and bacterial sugar transport systems are homologous. *Nature*, **325**, 641−643.

Markgraf,M., Bocklage,H. and Müller-Hill,B. (1985) A change of threonine 266 to isoleucine in the *lac* permease of *Escherichia coli* diminishes transport of lactose and increases transport of maltose. *Mol. Gen. Genet.*, **198**, 473−475.

McMurray,L., Petrucci,R.E. and Levy,S.B. (1980) Active efflux of tetracyclin encoded by four genetically different tetracyclin resistance determinants in *Escherichia coli. Proc. Natl. Acad. Sci. USA*, **77**, 3974–3977.

Messing,J., Crea,R. and Seeburg,P.H. (1981) A system for shotgun DNA sequencing. *Nucleic Acids Res.*, **9**, 309–321.

Michels,P.A.M., Michels,J.P.S., Boonstra,T. and Konings,W.N. (1979) Generation of an electrochemical proton gradient in bacteria by the excretion of metabolic end products. *FEMS Microbiol. Lett.*, **5**, 357–364.

Mitchell,P. (1963) Molecule, group and electron translocation through natural membranes. *Biochem. Soc. Symp.*, **22**, 142–169.

Mitchell,P. (1966) Chemiosmotic coupling in oxidative and photosynthetic phosphorylation. *Biol. Rev.*, **41**, 445–502.

Mitchell,P. (1973) Performance and conservation of osmotic work by proton-coupled solute porter systems. *Bioenergetics*, **4**, 63–91.

Mott,J.E., Grant,R.A., Ho,Y.-S. and Platt,T. (1985) Maximising gene expression from plasmid vectors containing the λP_L promoter: strategies for overproducing transcription termination factor ϱ. *Proc. Natl. Acad. Sci. USA*, **82**, 88–92.

Mueckler,M., Caruso,C., Baldwin,S.A., Panico,M., Blench,I., Morris,H.R., Allard,W.J., Lienhardt,G.E. and Lodish,H.F. (1985) Sequence and structure of a human glucose transporter. *Science*, **229**, 941–945.

Muir,M., Williams,L. and Ferenci,T. (1985) The influence of transport energisation on the growth yield of *Escherichia coli. J. Bacteriol.*, **163**, 1237–1242.

Nakae,T. (1986) Outer-membrane permeability of bacteria. *Crit. Rev. Microbiol.*, **13**, 1–62.

Neu,H.C. and Heppel,L.A. (1965) The release of enzymes from *Escherichia coli* during the formation of spheroplasts. *J. Biol. Chem.*, **240**, 3685–3692.

Newman,M.J., Foster,D.L., Wilson,T.H. and Kaback,H.R. (1981) Purification and reconstitution of functional lactose carrier from *Escherichia coli. J. Biol. Chem.*, **256**, 11804–11808.

Nikaido,H. (1986) Transport through the outer membrane of bacteria. *Methods Enzymol.*, **125**, 265–278.

Nikaido,H. and Rosenberg,E.Y. (1983) Porin channels in *Escherichia coli*: studies with liposomes reconstituted from purified proteins. *J. Bacteriol.*, **153**, 241–252.

Nikaido,H. and Vaara,M. (1985) Molecular basis of bacterial outer membrane permeability. *Microbiol. Rev.*, **49**, 1–32.

Okada,T., Veyama,K., Niiya,S., Kanazawa,H., Futai,M. and Tsuchiya,T. (1981) Role of inducer exclusion in preferential utilisation of glucose over melibiose in diauxic growth of *Escherichia coli. J. Bacteriol.*, **146**, 1030–1037.

Ordal,G.W. (1985) Bacterial chemotaxis. *Crit. Rev. Microbiol.*, **12**, 95–130.

Postma,P.W. and Lengeler,J.W. (1985) PEP:carbohydrate phosphotransferase system of bacteria. *Microbiol. Rev.*, **49**, 232–269.

Quiocho,F.A. (1986) Carbohydrate-binding proteins: tertiary structures and protein-sugar interactions. *Annu. Rev. Biochem.*, **55**, 287–315.

Quiocho,F.A. and Vyas,N.K. (1984) Novel stereospecificity of the L-arabinose binding protein. *Nature*, **310**, 381–386.

Raibaud,O. and Schwartz,M. (1980) Restriction map of the *Escherichia coli malA* region and identification of the *malT* product. *J. Bacteriol.*, **143**, 761–771.

Raibaud,O., Roa,M., Braun-Breton,C. and Schwartz,M. (1980) Genetic map of the *malK-lamB* operon. *Mol. Gen. Genet.*, **174**, 241–248.

Robbins,A.R., Guzman,R. and Rotman,B. (1976) Roles of individual *mgl* gene products in the β-methyl-galactoside transport system of *Escherichia coli* K12. *J. Biol. Chem.*, **251**, 3112–3116.

Rogers,H.J., Perkins,H.R. and Ward,J.B. (1980) *Microbial Cell Walls and Membranes*. Chapman and Hall, London.

Rosen,B.P. and Kashket,E.R. (1978) Energetics of active transport. In *Bacterial Transport*. Rosen,B.P. (ed.), Marcel Dekker Inc., New York, pp. 559–620.

Rosenberg,M., McKenney,K. and Schumperli,D. (1982) In *Promoters, Structure and Function*. Chambolin,M. and Rodriguez,M.R.L. (eds), Praeger, New York, pp. 380–387.

Saier,M.H. (1985) *Mechanisms and Regulation of Carbohydrate Transport in Bacteria*. Academic Press, New York.

Sanger,F., Coulsen,A.R., Barrell,B.G., Smith,A.J.H. and Roe,B.A. (1980) Cloning in single-stranded bacteriophage as an aid to rapid DNA sequencing. *J. Mol. Biol.*, **143**, 161–178.

Saper,M.A. and Quiocho,F.A. (1983) Leucine, isoleucine, valine-binding protein from *Escherichia coli* structure at 3 Å resolution and location of the binding site. *J. Biol. Chem.*, **258**, 11057–11062.

Sarkar,H.K., Viitanen,P.V., Trumble,W.R., Poonian,M.S., McComas,W. and Kaback,H.R. (1986) Oligo-nucleotide-directed site-specific mutagenesis of the *lac* permease of *Escherichia coli*. *Methods Enzymol.*, **125**, 214−229.

Schultz,S.G. and Curran,P.F. (1970) Coupled transport of sodium and organic solutes. *Physiol. Rev.*, **550**, 637−718.

Sasatsu,M., Misra,T.K., Chu,L., Ladagu,R. and Silver,S. (1985) Cloning and DNA sequence of a plasmid-determined citrate utilisation system in *Escherichia coli*. *J. Bacteriol.*, **164**, 983−993.

Shuman,H.A. and Silhavy,T.J. (1981) Identification of the *malK* gene product. *J. Biol. Chem.*, **256**, 560−562.

Shuman,H.A., Silhavy,T.J. and Beckwith,J.R. (1980) Identification of a cytoplasmic membrane component of the *Escherichia coli* maltose transport system. *J. Biol. Chem.*, **255**, 168−174.

Silhavy,T.J., Berman,H.L. and Enquist,L.W. (1984) *Experiments with Gene Fusions*. Cold Spring Harbor Laboratory Press, USA.

Silhavy,T.J., Ferenci,T. and Boos,W. (1978) Sugar transport systems in *Escherichia coli*. In *Bacterial Transport*. Rosen,B.P. (ed.), Marcel Dekker Inc., New York, pp. 127−169.

Stern,M.J., Ames,G.F.-L., Smith,N.H., Robinson,E.C. and Higgins,C.F. (1984) Repetitive extragenic palindromic sequences: a major component of the bacterial genome. *Cell*, **37**, 1015−1026.

Stock,J. and Roseman,S. (1971) A sodium-dependent sugar cotransport system in bacteria. *Biochem. Biophys. Res. Commun.*, **44**, 132−138.

Stoeckenius,W., Lozier,R. and Bogomoloni,R.A. (1979) Bacteriorhodopsin and the purple membrane of halobacteria. *Biochim. Biophys. Acta*, **505**, 215−278.

Stoner,C. and Schleif,R. (1983) The *araE* low affinity L-arabinose transport promoter. *J. Mol. Biol.*, **171**, 369−381.

Thompson,J. (1980) Galactose transport systems in *Streptococcus lactis*. *J. Bacteriol.*, **144**, 683−691.

Tokuda,H. (1986) Sodium translocation by NADH oxidase of *Vibrio alginolyticus*: isolation and characterisation of the sodium pump-defective mutants. *Methods Enzymol.*, **125**, 520−530. Academic Press, New York, Vol. 125, pp. 520−530.

Tsuchiya,T. and Wilson,T.H. (1978) Cation-sugar cotransport in the melibiose transport system of *Escherichia coli*. *Membr. Biochem.*, **2**, 63−79.

Tsuchiya,T., Takeda,K. and Wilson,T.H. (1980) H⁺-substrate cotransport by the melibiose membrane carrier in *Escherichia coli*. *Membr. Biochem.*, **3**, 131−146.

Tsuchiya,T., Raven,J. and Wilson,T.H. (1977) Co-transport of Na⁺ and methyl-β-D-thiogalactoside mediated by the melibiose transport system of *Escherichia coli*. *Biochem. Biophys. Res. Commun.*, **76**, 26−31.

Vyas,N.K., Vyas,M.N. and Quiocho,F.A. (1983) The 3 Å resolution structure of the D-galactose binding protein for transport and chemotaxis in *Escherichia coli*. *Proc. Natl. Acad. Sci. USA*, **80**, 1792−1796.

Walker,J.E., Saraste,M., Runswick,M.J. and Gay,N.J. (1982) Distantly related sequences in the α- and β-subunits of ATP synthase, myosin, kinases and other ATP-requiring enzymes and a common nucleotide binding fold. *EMBO J.*, **1**, 945−951.

Walker,J.E., Saraste,M. and Gay,N.J. (1984) Nucleotide sequence, regulation and structure of ATP-synthase. *Biochim. Biophys. Acta*, **768**, 164−200.

Waygood,E.B., Mattoo,R.L. and Peri,K.G. (1984) Phosphoproteins and the phosphoenol pyruvate:sugar phosphotransferase system in *Salmonella typhimurium* and *Escherichia coli*: evidence for III[Man], III[Fru], III[Glucitol], and the phosphorylation of EnzymeII[Mannitol], and EnzymeII[N-acetyl glucosamine]. *J. Cell. Biochem.*, **25**, 139−159.

West,I.C. (1970) Lactose coupled to proton movements in *Escherichia coli*. *Biochem. Biophys. Res. Commun.*, **41**, 655−661.

West,I.C. (1980) Energy coupling in secondary active transport. *Biochim. Biophys. Acta*, **604**, 91−126.

West,I.C. (1983) *The Biochemistry of Membrane Transport*. Chapman and Hall, London.

West,I.C. and Mitchell,P. (1972) Proton-coupled β-galactoside transport in non-metabolising *Escherichia coli*. *Bioenergetics*, **3**, 445−462.

West,I.C. and Mitchell,P. (1973) Stoicheiometry of lactose-proton symport across the plasma membrane of *Escherichia coli*. *Biochem. J.*, **132**, 587−592.

West,I.C. and Mitchell,P. (1974) Proton/sodium ion antiport in *Escherichia coli*. *Biochem. J.*, **144**, 87−90.

Wilson,D.B. (1974) The regulation and properties of the galactose transport system in *Escherichia coli* K12. *J. Biol. Chem.*, **249**, 553−558.

Wilson,D.M., Tsuchiya,T. and Wilson,T.H. (1986a) Methods for the study of the melibiose carrier of *Escherichia coli*. *Methods Enzymol.*, **125**, 377−386.

Winter,G., Fersht,A.R., Wilkinson,A.J., Zoller,M. and Smith,M. (1982) Redesigning enzyme structure by site-directed mutagenesis: tyrosyl-tRNA synthetase ATP binding. *Nature,* **299**, 756−758.

Yazyu,H., Shiota-Niiya,S., Shimamoto,T., Kanazawa,H., Futai,M. and Tsuchiya,T. (1983) Nucleotide sequence of the *melB* gene and characteristics of deduced amino acid sequence of the melibiose carrier in *Escherichia coli. J. Biol. Chem.,* **259**, 4320−4366.

Zoller,M.J. and Smith,M. (1983) Oligonucleotide-directed mutagenesis of DNA fragments cloned with M13 vectors. *Methods Enzymol.,* **100**, 468−500.

CHAPTER 6

The biochemistry of methane and methanol utilization

C.ANTHONY

Biochemistry Department, University of Southampton, UK

Introduction

The potential of a particular carbon substrate for use in biotechnology is sometimes merely a reflection of its cheapness or availability for use as a growth substrate; the biochemistry of its utilization may not have any special relevance or interest. This is not the case with methane and methanol. Although the relatively low cost and ready availability of these substrates is obviously relevant, the wide range of potential uses of methane and methanol is mainly a reflection of the diversity and novelty of the biochemistry of methylotrophs that use these substrates. It is for this reason that a chapter on the biochemistry of methane and methanol utilization is appropriate in a book on carbon substrates in biotechnology. Chapter 7 of this volume discusses the application of methylotrophs in biotechnology.

Methylotrophic microbes include yeasts and bacteria; these may be obligate methylotrophs or facultative, aerobic or anaerobic, Gram-positive or Gram-negative, non-motile, flagellate or stalked, with no internal compartments or endowed with extensive internal membrane systems or organelles; they may produce polysaccharides as components of capsules or as carbon stores, or they may store poly-β-hydroxybutyrate. This diversity of microbial type is matched by the diversity of biochemistry within the methylotrophs. There are four different specific pathways for assimilation of cell carbon, most of which have specific novel enzymes. Furthermore, the enzymes for the initial oxidation of methane and methanol to formaldehyde are all unusual, as are some aspects of the mechanism of energy transduction during methanol oxidation.

A knowledge of all these aspects of the biochemistry of methylotrophs is necessary in order to compare and evaluate particular methylotrophs for their use in biotechnology, whether it be for production of useful enzymes, for overproduction of metabolites or storage compounds or for growth of methylotrophs as a source of single cell protein. This chapter aims to provide a summary of our knowledge of the biochemistry of methane and methanol utilization; this has been based largely on previously published reviews (Anthony, 1980, 1982, 1986). The first section will deal with the main carbon assimilation pathways and their regulation. The second section will describe those unusual enzymes involved in catalysing methane and methanol oxidation, and the final section will deal with some aspects of energy transduction in methylotrophs.

Table 1. Summary of assimilation pathways in methylotrophs.

Pathway	Location	Comment	Summary equations for production of phosphoglycerate
1. RuBP pathway Calvin cycle (*Figure 1*)	Only in facultative autotrophs (e.g. *Paracoccus denitrificans*)	Rare in methylotrophs	$3CO_2 + 5NADH + 8ATP \rightarrow$ phosphoglycerate $+ 5NAD^+ + 8ADP + 7P_i$
2. RuMP pathway; KDPG aldolase/TA variant (*Figure 2a*)	In type I methanotrophs and in obligate methanol-utilizers	Requires Entner/Doudoroff enzymes and enzymes for conversion of pyruvate to triose phosphate but not phosphofructokinase or FBP aldolase	$3HCHO + NAD^+ + 2ATP \rightarrow$ phosphoglycerate $+ NADH + 2 ADP + P_i$
3. RuBP pathway: FBP aldolase/SBPase variant (*Figure 2b*)	In some facultative methylotrophs only some of which grow on methanol; not in methanotrophs	Requires phosphofructokinase and 'glycolytic' enzymes	$3HCHO + NAD^+ + ATP \rightarrow$ phosphoglycerate $+ NADH + ADP$
4. Serine pathway (*Figure 4*)	In facultative methanol-utilizers; in type II methanotrophs but not in other obligate methylotrophs	FPH_2 in reduced succinate dehydrogenase, produced during acetyl Co-A oxidation in icl^+ variant; an alternative reductant may be produced in icl^- variant	$2HCHO + CO_2 + 2NADH + 3ATP \rightarrow$ phosphoglycerate $+ 2NAD^+ + 3ADP + 2P_i + FPH_2$
5. DHA pathway (*Figure 5*)	Only in yeasts; not in bacteria	Fixation enzyme (DHA synthase) is in peroxisomes; rearrangement enzymes are in the cytosol	$3HCHO + NAD^+ + 2ATP \rightarrow$ phosphoglycerate $+ NADH + 2ADP + P_i$

The assimilation of methane and methanol

There are four diffierent pathways by which aerobic methylotrophs assimilate their carbon substrate into cell material and each has at least two potential variants; one route operates in yeast and the other three in bacteria. One pathway (the serine pathway) involves carboxylic acids and amino acids as intermediates whereas the other three all involve carbohydrate intermediates. *Table 1* summarizes the main assimilation pathways which are illustrated in full in *Figures, 1,2,4,* and *5*).

The ribulose bisphosphate pathway

The ribulose bisphosphate (RuBP) pathway, also called the Calvin cycle, for assimilation of carbon dioxide is the same as that occurring in plants and autotrophic bacteria (*Figure 1*). It only occurs in those methylotrophs that are also facultative autotrophs (e.g. *Paracoccus denitrificans*) and it does not occur in bacteria growing on methane, or in yeasts. It is the only carbon assimilation pathway that occurs in methylotrophs but is not peculiar to them. All the carbon is fixed at the level of carbon dioxide by way of RuBP carboxylase:

$$
\begin{array}{ccc}
\mathrm{CH_2OP} & & \\
| & & \\
\mathrm{C{=}O} & \mathrm{CO_2} & \mathrm{COOH} \\
| & & | \\
\mathrm{H\text{-}C\text{-}OH} & \longrightarrow & \mathrm{H\text{-}C\text{-}OH} \\
| & & | \\
\mathrm{H\text{-}C\text{-}OH} & \mathrm{H_2O} & \mathrm{CH_2OP} \\
| & & \\
\mathrm{CH_2OP} & & \\
\text{ribulose bisphosphate} & & \text{phosphoglycerate (2 mol)}
\end{array}
$$

Because of the highly oxidized nature of the intial substrate, the RuBP pathway requires relatively large amounts of NADH and ATP which is reflected in very low molar growth yields. Since this route is rare in methylotrophs and not peculiar to them this pathway will not be discussed further.

The ribulose monophosphate pathway

The ribulose monophosphate (RuMP) pathway operates only during methylotrophic growth of bacteria (not yeasts), fixing carbon at the oxidation level of formaldehyde, produced by the oxidation of methane or methanaol (*Table 1, Figure 2*). The overall reaction cycle synthesizes one molecule of pyruvate or DHA phosphate from three molecules of formaldehyde.

The first part of the pathway, catalysed by hexulose phosphate synthase, is the aldol condensation of formaldehyde with ribulose 5-phosphate to give 3-hexulose 6-phosphate:

$$
\begin{array}{ccc}
& & \mathrm{CH_2OH} \\
& \mathrm{CH_2OH} & \mathrm{HO\text{-}C\text{-}H} \\
& | & | \\
& \mathrm{C{=}O} & \mathrm{C{=}O} \\
& | & | \\
\mathrm{HCHO} \quad + & \mathrm{H\text{-}C\text{-}OH} \longrightarrow & \mathrm{H\text{-}C\text{-}OH} \\
& | & | \\
& \mathrm{H\text{-}C\text{-}OH} & \mathrm{H\text{-}C\text{-}OH} \\
& | & | \\
& \mathrm{CH_2O\text{-}P} & \mathrm{CH_2O\text{-}P}
\end{array}
$$

In methanotrophs hexulose phosphate synthase is large and membrane-bound, whereas

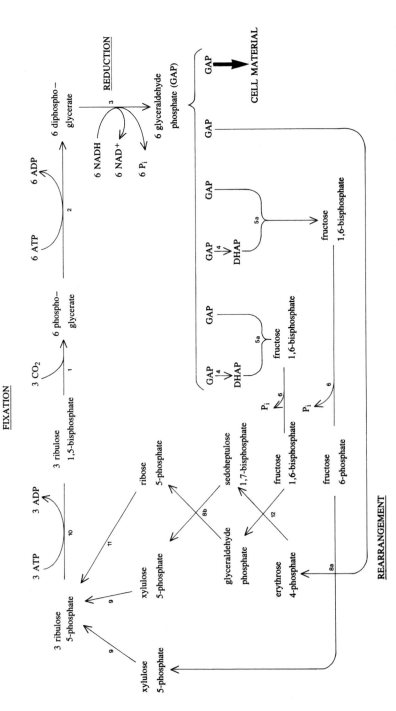

Figure 1. The RuBP cycle of CO_2 assimilation. (1) RuBP carboxylase; (2) phosphoglycerate kinase; (3) glyceraldehyde phosphate dehydrogenase; (4) triose phosphate isomerase; (5) aldolase; (6) fructose bisphosphatase; (8) transketolase; (9) pentose phosphate epimerase; (10) phosphoribulokinase; (11) pentose phosphate isomerase; (12) transaldolase. Reproduced from Anthony (1982) with permission from Academic Press.

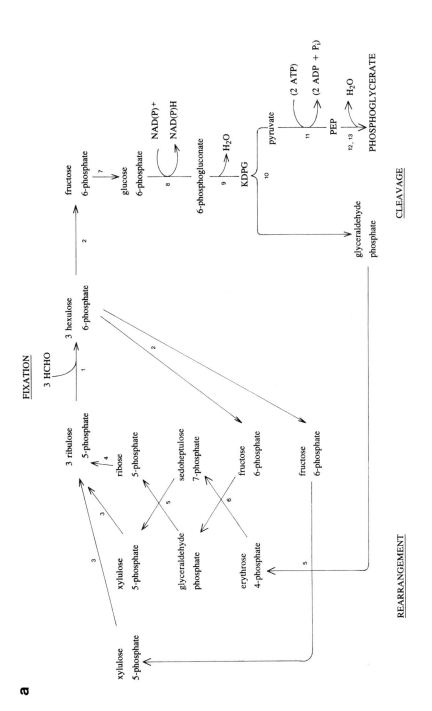

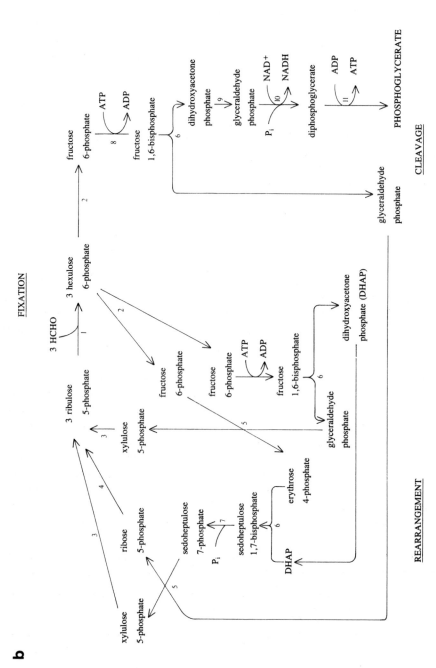

in methanol-utilizers it is a soluble dimeric enzyme with a total molecular weight of about 40 000 daltons.

In the second part of the pathway a C-6 compound is cleaved by way of an aldolase to two C-3 compounds. In one variant of the pathway 2-keto 3-deoxy 6-phosphogluconate (KDPG) is cleaved to glyceraldehyde 3-phosphate plus pyruvate; in the other variant fructose bisphosphate is cleaved to glyceraldehyde 3-phosphate plus dihydroxyacetone phosphate.

In the final part of the pathway the glyceraldehyde phosphate plus two molecules of fructose phosphate undergo rearrangement reactions to regenerate the acceptor molecule of the pathway, ribulose 5-phosphate.

Because there are two possible cleavage enzymes and two rearrangement sequences there are four potential variants of the RuMP pathway but only two of these are of any importance. The KDPG aldolase/transaldolase variant occurs in obligate methylotrophs growing on methane or methanol; these bacteria include *Methylococcus capsulatus* and *Methylomonas methanica* (both type I methanotrophs) and *Methylophilus methylotrophus*. The main alternative, the fructose bisphosphate (FBP) aldolase/sedoheptulose (SBP) variant, occurs in some facultative methylotrophs, including *Bacillus* and *Arthrobacter* species, only a few of which grow on methanol and none of which grows on methane.

It should be noted that because it is essential to produce both triose phosphate and pyruvate for biosynthesis, bacteria with the KDPG aldolase variant must contain the 'gluconeogenic' enzymes for synthesis of triose phosphate from pyruvate. Similarly, bacteria with the fructose bisphosphatase variant must have all the 'glycolytic' enzymes for production of pyruvate from triose phosphate.

Some bacteria having the RuMP pathway for assimilation of formaldehyde are able to oxidize formaldehyde by a variant of the assimilation pathway (*Figure 3*). These bacteria must also contain glucose 6-phosphate dehydrogenase and 6-phosphogluconate dehydrogenase. Oddly enough, not all those bacteria having the KDPG aldolase variant (and hence glucose 6-phosphate dehydrogenase) have 6-phosphogluconate dehydrogenase. Obligate methanol-utilizers such as *Methylophilus methylotrophus* do have such a dissimilatory cycle and these bacteria have low levels of enzymes able directly to oxidize formaldehyde and formate. By contrast, in the methanotrophic bacteria the oxidation of these substrates is probably by way of NAD-linked dehydrogenases, and not the dissimilatory cycle. Even when these bacteria (such as *M. capsulatus*) have glucose phosphate dehydrogenase and the KDPG variant of the assimilation pathway,

Figure 2. (a) The RuMP pathway of formaldehyde assimilation (KDPG aldolase/transaldolase variant). This variant occurs only in obligate methylotrophs. (1) Hexulose phosphate synthase; (2) hexulose phosphate isomerase; (3) pentose phosphate epimerase; (4) pentose phosphate isomerase; (5) transketolase; (6) transaldolase; (7) glucose phosphate isomerase; (8) glucose phosphate dehydrogenase; (9) phosphogluconate dehydrase; (10) KDPG aldolase; (11) PEP synthetase or equivalent enzymes; (12) enolase; (13) phosphoglyceromutase.
(b) The RuMP pathway of formaldehyde assimilation (fructose bisphosphate aldolase/sedoheptulose bisphosphatase variant). This variant occurs only in facultative methylotrophs. (1) Hexulose phosphate synthase; (2) hexulose phosphate isomerase; (3) pentose phosphate epimerase; (4) pentose phosphate isomerase; (5) transketolase; (6) aldolase; (7) sedoheptulose bisphosphatase; (8) phosphofructokinase; (9) triose phosphate isomerase; (10) glyceraldehyde phosphate dehydrogenase; (11) phosphoglycerate kinase. Reproduced from Anthony (1982) with permission from Academic Press.

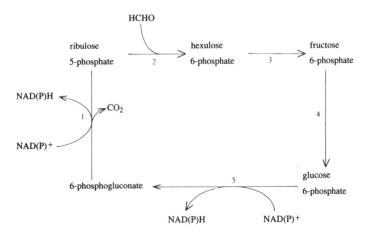

Figure 3. The oxidation of formaldehyde to CO_2 by a dissimilatory RuMP cycle. The enzymes of this cycle are those of the assimilatory cycle plus the dissimilatory enzyme 6-phosphogluconate dehydrogenase (reaction **1**). Other enzymes: **(2)** hexulose phosphate synthase; **(3)** hexulose phosphate isomerase; **(4)** glucose phosphate isomerase; **(5)** glucose 6-phosphate dehydrogenase.

they usually have low levels of phosphogluconate dehydrogenase (Zatman, 1981).

The problem of balancing the flow of formaldehyde into assimilation and oxidation in bacteria with the RuMP pathway must be solved differently in those bacteria having the cyclic dissimilatory route and those oxidizing formaldehyde by way of formaldehyde and formate dehydrogenases. It so happens that the bacteria using this direct route for oxidation of formaldehyde are the methanotrophs. These differ from the methanol-utilizers in having extensive internal membrane systems (in some conditions) and in having a membrane-bound hexulose phosphate synthase. Almost nothing is known about regulation of formaldehyde metabolism in these bacteria.

In most other obligate methylotrophs (methanol-utilizers) formaldehyde is oxidized by the cyclic dissimilatory route. The hexulose phosphate synthase and isomerase have dual roles (assimilation and oxidation), and hence neither these enzymes nor the reversible rearrangement enzymes are likely to be involved in regulation. In these organisms the branch point of formaldehyde metabolism is at the level of 6-phosphogluconate. Glucose phosphate dehydrogenase is involved in both biosynthesis and oxidation whereas phosphogluconate dehydrogenase is involved only in the oxidative cycle. Both these dehydrogenases are inhibited by ATP and NAD(P)H but the extent of inhibition is different. The phosphogluconate dehydrogenase will need to be inhibited to a greater extent than the glucose phosphate dehydrogenase because this enzyme is also involved in assimilation (see Ben-Bassat and Goldberg, 1980; Beardsmore *et al.*, 1982).

The serine pathway

The serine pathway differs from other C-1 assimilation pathways in the nature of its intermediates which are carboxylic acids and amino acids rather than carbohydrates. The possession of high concentrations of some of the enzymes necessary for the interconversion of these molecules may make bacteria having this pathway suitable for manipulation to give overproduction of some of them. Carbon is fixed into this pathway in two reactions, the first being catalysed by serine transhydroxymethylase:

$$\text{HCHO} + \underset{\overset{|}{\text{COOH}}}{\text{CH}_2\text{NH}_2} \longrightarrow \underset{\overset{|}{\underset{|}{\text{COOH}}}}{\overset{\text{CH}_2\text{OH}}{\text{CHNH}_2}}$$

The second 'fixation' reaction is catalysed by phosphoenolpyruvate (PEP) carboxylase, the product being oxaloacetate:

$$\underset{\underset{\text{COOH}}{|}}{\overset{\overset{\text{CH}_2}{\|}}{\text{C-OP}}} + \text{CO}_2 \longrightarrow \underset{\underset{\text{COOH}}{|}}{\overset{\overset{\text{CH}_2\text{COOH}}{|}}{\text{C}=\text{O}}} + \text{P}_i$$

In *Figure 4* the serine pathway is drawn as it operates to synthesize phosphoglycerate from two molecules of formaldehyde plus one of carbon dioxide. In the first part of the pathway two molecules of formaldehyde plus two of glyoxylate give two molecules of 2-phosphoglycerate. One of these is assimilated into cell material by way of 3-phosphoglycerate while the other is converted by a series of reactions to malyl-CoA. In the final part of the pathway this is cleaved by malyl-CoA lyase to glyoxylate and acetyl-CoA whose oxidation to a second molecule of glyoxylate complete the cycle. In the icl$^+$-serine pathway the acetyl-CoA is oxidized by a glyoxylate cycle that involves isocitrate lyase. In the icl$^-$-serine pathway (as found for example in *Peudomonas* AM1) isocitrate lyase is absent but no alternative route is known for the oxidation of acetyl-CoA to glyoxylate. One alternative route that has been proposed (the homoisocitrate cycle) appears to be a figment of the imagination. The serine pathway can operate to produce C-4 carboxylic acids without using any additional enzymes, but for biosynthesis of carbohydrates from phosphoglycerate the usual gluconeogenic enzymes are required, together with pyruvate kinase for the production of pyruvate from PEP. The route for production of acetyl-CoA involves malyl-CoA lyase, this being the only possible route for those bacteria such as obligate methylotrophs and the *Hyphomicrobia* which lack pyruvate dehydrogenase.

Most serine pathway bacteria are facultative methylotrophs; the only obligate methylotrophs having the serine pathway are the type II methanotrophic bacteria such as *Methylosinus trichosporium*. The only bacteria known to have the icl$^+$ variant of the serine pathway are Gram-negative facultative methylotrophs most of which are unable to use methane or methanol (they use methylated amines).

Because the serine pathway occurs in facultative methylotrophs it is possible to investigate the regulation of synthesis of the serine pathway enzymes at the genetic level but this topic is outside the limits of this review (see Anthony, 1982; Fulton *et al.*, 1984; Tatra and Goodwin, 1983, 1984, 1985).

The problem of regulation of activity of the serine pathway is similar to that in the RuMP pathway. The first requirement for regulation is that the proportion of formaldehyde that is oxidized or assimilated must be controlled. By contrast with the RuMP pathway, free formaldehyde is not the substrate for assimilation by the serine pathway; formaldehyde is first converted to methylenetetrahydrofolate which is the substrate for serine transhydroxymethylase. This reaction can occur non-enzymically and is unlikely to be a site of regulation. The only enzymes so far identified as potential regulatory

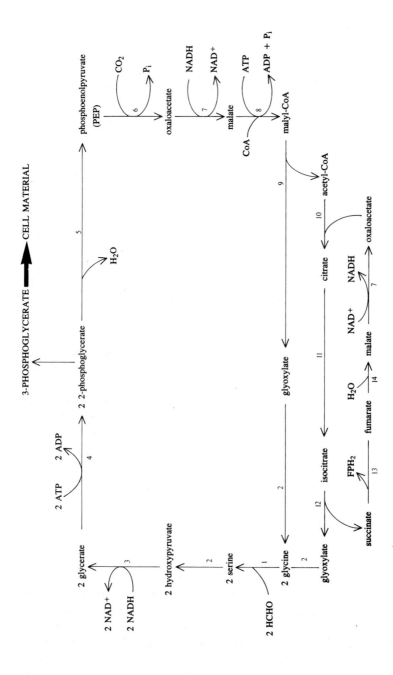

Figure 4. The serine pathway (icl⁻ variant). The icl⁻ variant differs from this in lacking measurable malate thiokinase and in having an alternative route for the oxidation of acetyl-CoA to glyoxylate not involving isocitrate lyase. Precursors for biosynthesis can be removed from the cycle at the level of oxaloacetate or succinate. (1) Serine transhydroxymethylase; (2) serine-glyoxylate aminotransferase; (3) hydroxypyruvate reductase; (4) glycerate kinase; (5) enolase; (6) PEP carboxylase; (7) malate dehydrogenase; (8) malate thiokinase; (9) malyl-CoA lyase; (10) citrate synthase; (11) aconitase; (12) isocitrate lyase; (13) succinate dehydrogenase; (14) fumarase. Reproduced from Anthony (1982) with permission from Academic Press.

enzymes in the serine pathway are the transhydroxymethylase and the PEP carboxylase, both of which are fixation enzymes in this pathway.

In those methylotrophs studied, there are two serine transhydroxymethylases, one of which is specifically produced during growth on C-1 substrates. This enzyme is activated by glyoxylate. Glyoxylate is thus acting as a measure of biosynthetic potential in the bacteria, stimulating the operation of the serine pathway.

A second enzyme involved in the regulation of the serine pathway in some methylotrophs (for example, *Pseudomonas* MA) is PEP carboxylase. This enzyme is induced during methylotrophic growth and is usually the only enzyme with appreciable activity in carboxylating C-3 compounds to C-4 compounds. In other organisms this enzyme is stimulated by acetyl-CoA, thus reflecting its role as an anaplerotic enzyme, replenishing the tricarboxylic acid cycle with C-4 substrates when required. The PEP carboxylase of *Pseudomonas* MA is not regulated in this way. It is activated by NADH which stimulates conversion of inactive tetramer to active dimeric enzyme. ADP is an allosteric inhibitor of the enzyme, producing inactive tetramers from active dimers. When concentrations of NADH and ATP are high (and ADP is low) then there is sufficient energy available for biosynthesis and the PEP carboxylase will have a high activity. Conversely when the concentration of NADH is low and ADP is high the rate of operation of the serine pathway will be diminished, thus increasing the relative flow of formaldehyde into catabolism for production of 'energy'. This control mechanism amplifies the general regulatory effects of the ratios of NADH/NAD and ATP/ADP.

The dihydroxyacetone pathway

The dihydroxyacetone (DHA) pathway, also called the xylulose monophosphate cycle, occurs only in yeasts (*Figure 5*). The only enzyme required exclusively for formaldehyde assimilation is the initial formaldehyde-fixing enzyme, DHA synthase; all other enzymes of the pathway are also required for growth on carbohydrates, ethanol or glycerol.

In the fixation part of the pathway DHA synthase catalyses the transfer of a glycolaldehyde moiety from xylulose 5-phosphate to formaldehyde, giving glyceraldehyde 3-phosphate and DHA:

$$
\begin{array}{c}
CH_2OH \\
| \\
C=0 \\
| \\
HO\text{-}C\text{-}H \\
| \\
HCHO + \quad H\text{-}C\text{-}OH \longrightarrow \\
| \\
CH_2O\text{-}P
\end{array}
\qquad
\begin{array}{c}
CHO \\
| \\
H\text{-}C\text{-}OH \\
| \\
CH_2O\text{-}P
\end{array}
+
\begin{array}{c}
CH_2OH \\
| \\
C=0 \\
| \\
CH_2OH
\end{array}
$$

This enzyme is induced during growth on methanol and is located in the peroxisomes together with the enzymes for oxidation of methanol to formaldehyde (see *Figure 7*) (Goodman *et al.*, 1984; Goodman, 1985; Douma *et al.*, 1985). The localization of DHA synthase in the peroxisome confers protection against inhibitory glutathione (GSH).

Dihydroxyacetone produced by DHA synthase passes out of the peroxisome into the cytoplasm where it is phosphorylated by a specific, inducible, triokinase to dihydroxyacetone phosphate. This is condensed with glyceraldehyde phosphate in an aldolase reaction, giving fructose 1,6-bisphosphate which is hydrolysed to fructose 6-phosphate. This is then rearranged with glyceraldehyde phosphate to regenerate xylulose

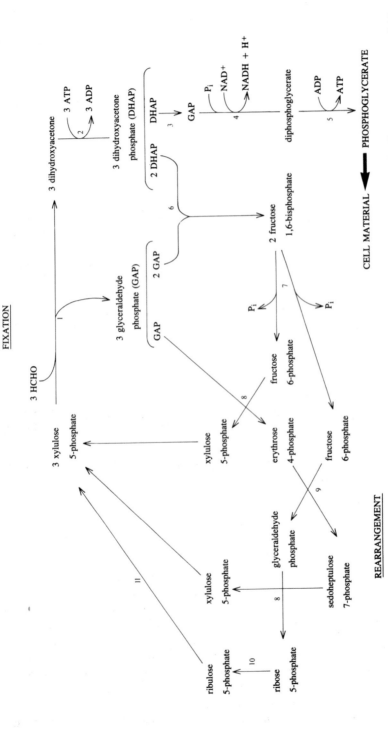

Figure 5. The DHA pathway of formaldehyde assimilation in yeast. This pathway is also called the xylulose monophosphate cycle. The rearrangement part of this cycle involves transaldolase (reaction **9**) but an alternative rearrangement might involve sedoheptulose bisphosphate aldolase and phosphatase instead of transaldolase (as in *Figure 2b*). (**1**) Dihydroxyacetone synthase; (**2**) triokinase; (**3**) triose phosphate isomerase; (**4**) glyceraldehyde phosphate dehydrogenase; (**5**) phosphoglycerate kinase; (**6**) fructose bisphosphate aldolase; (**7**) fructose bisphosphatase; (**8**) transketolase; (**9**) transaldolase; (**10**) pentose phosphate isomerase; (**11**) pentose phosphate epimerase. Reproduced from Anthony (1982) with permission from Academic Press.

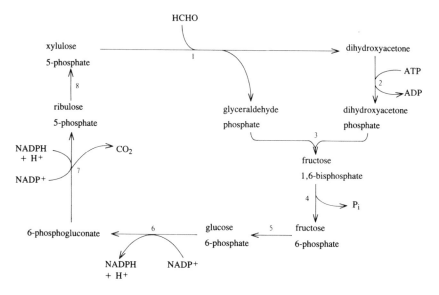

Figure 6. The dissimilatory DHA pathway of formaldehyde oxidation in yeast. (**1**) Dihydroxyacetone synthase; (**2**) triokinase; (**3**) fructose bisphosphate aldolase; (**4**) fructose bisphosphatase; (**5**) glucose phosphate isomerase; (**6**) glucose phosphate dehydrogenase; (**7**) 6-phosphogluconate dehydrogenase; (**8**) pentose phosphate epimerase. Reproduced from Anthony (1982) with permission from Academic Press.

monophosphate which must enter the peroxisome for a further round of formaldehyde fixation.

For production of PEP, pyruvate, acetyl-CoA and C-4 carboxylic acids from DHA phosphate the 'glycolytic' enzymes for oxidation of triose phosphate are also required, together with fructose bisphosphate aldolase and phosphatase which act in the opposite 'gluconeogenic' direction.

In the same way as some bacteria have modified RuMP pathway which acts as a dissimilatory cycle for formaldehyde oxidation (*Figure 3*), yeasts have a dissimilatory DHA cycle able to oxidize formaldehyde to carbon dioxide (*Figure 6*). By contrast with the bacterial cycle, the dehydrogenases in this cycle are specific for NADP and it has been concluded that the dissimilatory cycle in yeasts functions only for the production of NADPH for biosynthesis and that it has no role in the provision of ATP. This cycle is regulated by way of its two dehydrogenases which are inhibited by the 'end-product' of the cycle, NADPH.

The peroxisomal location of the DHA synthase may be involved in the regulation of formaldehyde assimilation. It has been suggested that the distribution of formaldehyde between the assimilatory and dissimilatory pathways might be controlled by the rate of transport of xylulose phosphate into the peroxisomes (Douma *et al.*, 1985). Regeneration of this sugar requires ATP (in the triokinase reaction) whose production will depend on the energy status of the cells. Metabolic control may act in such a way that in conditions of intracellular energy limitation the rate of supply of xylulose phosphate to the DHA synthase is diminished, thereby resulting in an increase in the amount of formaldehyde available for oxidation.

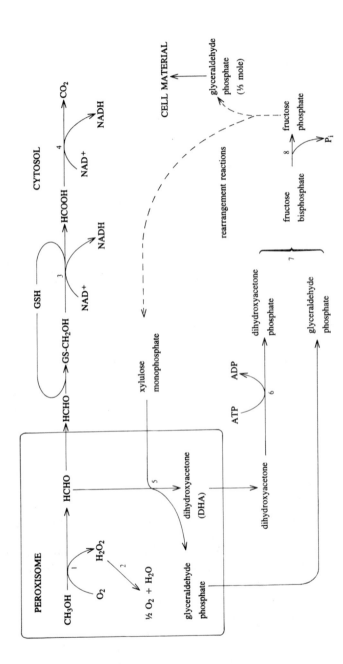

Figure 7. Compartmentation of oxidation and assimilation of methanol in yeast. (1) alcohol oxidase; (2) catalase; (3) glutathione$^+$ (GSH)-dependent formaldehyde dehydrogenase; (4) formate dehydrogenase; (5) dihydroxyacetone synthase; (6) dihydroxyacetone kinase (triokinase); (7) fructose bisphosphate aldolase; (8) fructose bisphosphatase.

The oxidation of methanol by yeasts

The oxidative metabolism of yeasts differs from that in bacteria in many ways including the nature and compartmentation of the oxidizing enzymes and the involvement of mitochondria in energy transduction (for reviews of the metabolism of methylotrophic yeasts see: Sahm, 1977; Tani *et al.*, 1979; Anthony, 1982; Veenhuis *et al.*, 1983). The overall process of methanol oxidation to carbon dioxide and its relation to the assimilation pathway is shown in *Figure 7*. The overall reaction is:

$$CH_2OH \ + \ 0.5 \ O_2 \ + \ 2 \ NAD^+ \longrightarrow CO_2 \ + \ 2 \ NADH$$

For oxidative phosphorylation to occur the NADH must enter the mitrochondria or it must be oxidized by an external NADH dehydrogenase. In either case, only two ATP molecules will be produced per molecule of NADH oxidized.

Methanol is oxidized to formaldehyde by a non-specific flavoprotein alcohol oxidase, molecular oxygen being consumed and hydrogen peroxide being produced. The oxidase has a total molecular weight of about 600 000 daltons and consists of eight subunits, each containing one molecule of non-covalently-bound FAD. The enzyme oxidizes only short-chain aliphatic primary alcohols which may be unsaturated or substituted. The K_m value for methanol of the oxidase (0.2−2.0 mM) is 10−100 times that measured for bacterial methanol dehydrogenase and it is similar to this enzyme in being able to oxidize formaldehyde. This oxidation is not important *in vivo* but it is not known how it is prevented. The K_m for oxygen is similar to the concentration of oxygen in air-saturated water (~0.2 mM) and this characteristic may be important in limiting the growth rate of yeasts in methanol-limited continuous culture. None of the energy available from the first step in the oxidation of methanol is harnessed as useable reductant (NADH) or as ATP and this will affect growth yields with respect to methanol and oxygen.

The oxidase exists in the yeast cells within peroxisomes whose function is to protect the cell from potential damage by peroxide formed as a result of oxidase activity. The peroxisomes are bounded by a unit membrane and they contain crystalloids consisting of crystalline alcohol oxidase. They also contain the non-crystalline catalase whose function is to remove the peroxide produced by the oxidase.

The formaldeyhyde produced by the action of alcohol oxidase is the substrate for the first assimilation enzyme, DHA synthase, which is also located in the peroxisome (see above). An alternative fate of the formaldehyde is oxidation, after leaving the peroxisome, to carbon dioxide and two molecules of NADH. The substrate for the cytosolic formaldehyde dehydrogenase is the thiohemiacetal formed spontaneously from reduced glutathione and formaldehyde (S-hydroxymethylglutathione). This is then the substrate for formate dehydrogenase, or it is first hydrolysed by a specific hydrolase to formate which is then oxidized by formate dehydrogenase.

The bacterial oxidation of methane and methanol
Introduction

The oxidative characteristics of yeasts were dealt with separately from those of bacteria because the yeasts do not use methane and their route for oxidation of methanol is completely different from that described here for bacteria. The bacterial oxidation of methane and methanol is described in more detail in recent reviews (Anthony, 1986; Dalton and Leak, 1985) which contain complete references to the original work.

Bacteria oxidizing methane and methanol do so by the following route:

$$CH_4 \xrightarrow[\text{NADH}]{O_2 \quad H_2O} CH_3OH \xrightarrow[\text{PQQH}_2]{} HCHO \xrightarrow[\text{'2H'}]{\quad \text{2NADH} \quad} HCOOH \xrightarrow[\text{NADH}]{} CO_2$$
$$\text{(HCOOH} \dashrightarrow CO_2\text{)}$$

The first reaction in methane oxidation is a hydroxylation catalysed by methane mono-oxygenase which requires a reductant that is probably always NADH. The oxidation of methanol to formaldehyde is always catalysed by methanol dehydrogenase (MDH) which has pyrrolo-quinoline quinone (PQQ) as its prosthetic group. The oxidation of formaldehyde to carbon dioxide must be coupled to the production of two molecules of NADH during growth on methane because one molecule must be used in the initial hydroxylation reaction. During growth on methanol, formaldehyde may be oxidized by alternative dehydrogenases that are not coupled to NADH production. Formate is always oxidized by an NAD-dependent dehydrogenase. Methanol-utilizers having the RuMP pathway for carbon assimilation have only very low levels of this dehydrogenase and formaldehyde oxidation is catalysed by a dissimilatory modification of the assimilation pathway. This pathway probably does not occur in methanotrophs.

The oxidation of methane to methanol

The nature of the first step in the oxidation of methane has many repercussions, both in terms of cell yields and in the potential use of methanotrophs for the controlled oxidation of other hydrocarbons.

The initial oxidation of methane is catalysed by a mixed function methane mono-oxygenase (MMO) that uses molecular oxygen together with a reductant. For some years it was thought that there were two different types of MMO, a soluble MMO using NADH as reductant and a membrane-bound MMO that used cytochrome c as reductant. Much of the initial confusion has now been resolved. There are two types of MMO, both using NADH as reductant. Particulate MMO is formed in conditions of copper sufficiency whereas soluble MMO is formed when bacteria are grown under conditions of copper insufficiency. Copper sufficiency and insufficiency are governed by copper concentrations in the growth medium and by the cell density; they can therefore appear to be determined by alterations in the growth rate or in the carbon, oxygen or nitrogen supply. A well-known feature of methanotrophic bacteria is their possession of intracytoplasmic membranes, and the physiological conditions affecting the formation of the two types of MMO also affect the formation of these membranes. Particulate MMO tends to be produced when extensive formation of membranes occurs, whereas the soluble form of MMO is the only form of the enzyme produced when bacteria are grown in conditions in which the internal membranes are fewer and less well-organized. It is not certain if the induction of the particulate MMO in copper-sufficient conditions leads to an increase in membrane content or vice versa; the two events may even be completely independent (Prior and Dalton, 1985).

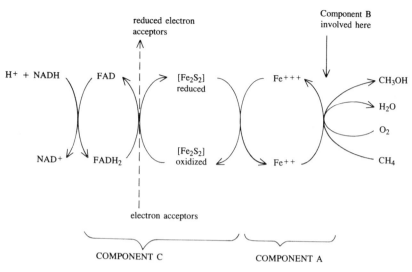

Figure 8. Pathway of electron transfer between the components of the soluble MMO complex during the oxidation of methane to methanol. This figure is based on the work of Professor H.Dalton and his colleagues.

It has been pointed out (Dalton and Leak, 1985) that two features must be borne in mind in considering the roles of the soluble and particulate systems. The first is that production of soluble MMO in conditions of low copper concentrations may represent a selective advantage to the organism able to adapt in this way (not all methanotrophs can do so). The second feature to consider is that in those organisms able to synthesize both systems the particulate system is made preferentially. They suggest that this preference is related to a greater growth efficiency in bacteria with the particulate MMO. It should be noted in this context, however, that it is difficult to envisage how efficiency might be selected or how it may confer an advantage. Furthermore, the higher carbon conversion efficiencies measured are in conditions where the membrane systems are fully developed and this may have an effect separate from the possession of particulate MMO.

(i) *The soluble methane mono-oxygenase.* After breakage of bacteria grown in conditions of copper insufficiency the MMO exists free in solution and consists of three components (A,B and C) which have been resolved and characterized by Dalton and his group at Warwick University (*Figure 8*) (see recent references: Dalton and Leak, 1985; Green and Dalton, 1985; Lund and Dalton, 1985; Lund *et al.*, 1985; Woodland and Dalton, 1984; Woodland and Cammack, 1985).

Component A is a large protein made up of two copies each of three different subunits. It is probable that component A is responsible for binding carbon substrate and oxygen during catalysis by the complete MMO system. It contains two atoms of non-haem iron per molecule but no acid-labile sulphur. Component A does not resemble other oxygenases; it is suggested that the unusual e.s.r. spectrum, the presence of iron and zinc, and the absence of haem, together with the apparent lack of an extrudable iron−sulphur cluster, indicate a novel iron-containing prosthetic group.

The site of binding and activation of oxygen is not known but it is probably first converted into an electron-deficient, metal-bound species which then reacts with the methane; the non-haem iron is clearly a likely candidate for the metal involved.

Component C is an iron–sulphur flavoprotein with a single polypeptide chain of molecular weight 44 000 daltons. Each molecule contains one molecule of FAD and a single iron–sulphur centre of the [2Fe-2S*(S-Cys)$_4$] type as found in spinach ferredoxin, and rubredoxin. These properties suggest that the single protein has a function analogous to the combination of putidaredoxin plus NADH-putidaredoxin reductase (a flavoprotein) in the hydroxylation of camphor. The iron–sulphur centre of component C is not essential for its NADH-acceptor reductase activity; it is only necessary for electron transfer to component A. The affinity of component C for NADH is much greater than its affinity for NADPH, suggesting that NADH is the natural electron donor.

Component B is a small colourless protein, having a single polypeptide chain and lacking any observable prosthetic group. Although not essential for the electron flow in the MMO from NADH to oxygen catalysed by component A, this flow is modified by component B. In its absence, components A plus C catalyse an 'NADH oxidase' reaction; whereas in the presence of component B the reaction of the hydroxylase (component A) with methane is facilitated and all electron flow is diverted to the oxygenase reaction. Thus besides being essential for oxygenase function, an important second role of component B is that of preventing 'wasteful' oxidation of NADH by 'NADH oxidase' activity of MMO in the absence of methane (Green and Dalton, 1985).

The soluble MMO is remarkably insensitive to chelating agents, thiol reagents and elctron transport inhibitors; this is especially so when compared with the particulate form of the enzyme. The only potent inhibitors are 8-hydroxyquinoline and acetylenic compounds. Other inhibitors that inhibit activity of the soluble MMO in whole cells probably do so by inhibiting the supply of NADH to the enzyme.

The substrate specificity of soluble MMO is different from that of particulate MMO. This is seen in whole bacteria and in the isolated enzymes. The soluble MMO is very non-specific. It catalyses the hydroxylation of primary and secondary alkyl C-H bonds, the formation of epoxides from internal and terminal alkenes, the hydroxylation of cyclohexane and aromatic compounds, the N-oxidation of pyridine, the oxidation of CO to carbon dioxide and the oxidation of methanol to formaldehyde. The only substrates oxidized by the MMO in whole cells, however, are those whose further oxidation yields NADH which is required for the initial hydroxylation reaction. Besides these substrates some further substrates can be oxidized by the MMO if an exogenous supply of reductant such as formaldehyde is also provided.

(ii) *The particulate methane mono-oxygenase*. Although a distinct, particulate MMO is certainly formed in methanotrophs during growth in conditions of copper sufficiency, little is known of its nature. One of the first descriptions of an MMO was of a system solubilized from membranes. This consisted of three proteins, one of which was an iron-containing protein, another being soluble cytochrome *c*.

The electron donor was either MDH or ascorbate. However, recent attempts to repeat the purification of the mono-oxygenase components or to confirm the findings with respect to electron donor have all been unsuccessful (see Higgins *et al.*, 1981; Dalton *et al.*, 1984).

When bacteria are transferred to a medium with sufficient copper, the protein components of the soluble system markedly diminish together with soluble MMO activity. Particulate MMO activity soon appears and three completely new proteins are synthesized, as shown by electrophoresis of particulate fractions, but as yet there is no evidence that these are components of MMO. It appears to be unlikely that the availability of copper merely alters the location of an otherwise identical MMO.

The conclusion that the particulate MMO is completely distinct from the soluble form is supported by the demonstration that the sensitivity to inhibitors changes with its location. The particulate MMO is inhibited by not only 8-hydroxyquinoline and acetylenic compounds (as is the soluble enzyme) but also by cyanide, mercaptoethanol, 2,2-dipyridyl and thiourea. A further indication that the particulate MMO is different from the soluble enzyme is its substrate specificity; it oxidizes fewer compounds than does the soluble enzyme, being unable, for example, to oxidize higher alkanes, cyclohexane and aromatic compounds.

(iii) *The electron donor for methane mono-oxygenase.* That the nature of the electron donor for MMO is likely to have a major influence on cell yields arises from the fact that every molecule of methane that is metabolized, either by assimilation to cell material or by oxidation to carbon dioxide, must consume one reducing equivalent. The reduction level of methane is thus 'energetically' equivalent to that of formaldehyde. The problem is, however, greater than this because the oxidation of methanol to formaldehyde (the assimilation substrate) does not usually provide NADH, the usual reductant for the MMO. It is for this reason that growth yields of methanotrophs can be said to be NADH-limited rather than ATP-limited. Increasing the P/O ratio in these bacteria would have a negligible effect on growth yields. These might be higher if the reductant for MMO were to be reduced MDH or reductant produced from it by 'reverse electron transport' (see Anthony, 1978, 1980, 1982, 1986; Dalton and Leak, 1985).

Although all forms of MMO require NADH as electron donor when measured in extracts (as described above), the situation may be different in whole organisms. The first alternative is that the NADH may be produced, not directly by NAD^+-dependent dehydrogenases but by 'reverse electron transport', from methanol by way of methanol dehydrogenase, cytochrome c and a low potential electron transport chain involving b-type cytochromes, quinones and iron−sulphur proteins. If this is the case then methane oxidation will be inhibited by inhibitors of this system. A second alternative is that NADH may not be the electron donor *in vivo* and that a component of a membrane-bound electron transport chain, such as an iron−sulphur protein, may be the immediate electron donor, the electrons arising by way of 'reverse electron transport' from MDH. There is some evidence that methanol or ethanol may be able to support hydroxylation of methane in *M. capsulatus* containing the particulate MMO but not the soluble MMO and that this permits a higher carbon conversion efficiency (Dalton and Leak, 1985; Leak *et al.*, 1985).

The oxidation of methanol to formaldehyde

(i) *Methanol dehydrogenase.* Methanol oxidation in bacteria is usually oxidized by NAD^+-independent MDH, possession of this enzyme being a feature that appears to be common to almost all bacteria able to grow on methane and methanol. The enzyme

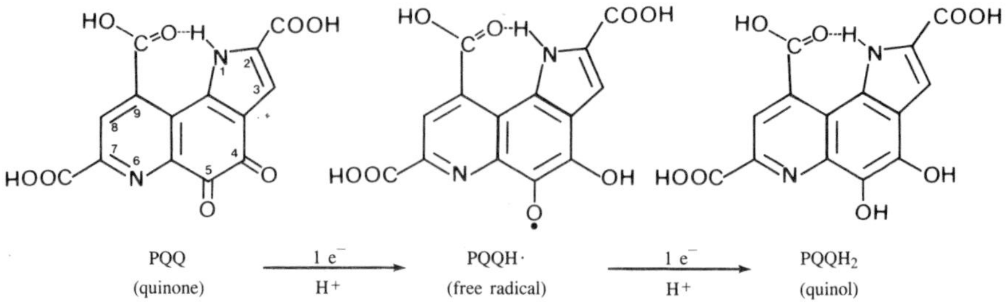

Figure 9. The prosthetic group of methanol dehydrogenase (PQQ). The full name of PQQ is 2,7,9-tricarboxy-1H-pyrrolo[2,3-f]quinoline-4,5-dione. An alternative trivial name is methoxatin.

is assayed *in vitro* at its pH optimum of about pH 9, in the presence of ammonia activator and the artificial electron acceptor, phenazine methosulphate or ethosulphate. Most MDHs are dimers of identical subunit molecules of molecular weight 60 000 daltons and most are basic proteins. Exceptions are the MDH from *Methylomonas methanica* which is a monomeric enzyme and the MDH from *Paracoccus denitrificans* which is an acidic protein.

MDH oxidizes a wide range of primary alcohols, the rates with most substrates being similar to those measured with methanol. By contrast with MMO, whole bacteria oxidize most of the substrates that are oxidized by the pure dehydrogenase. This is partly because the dehydrogenase is readily accessible to most substrates, being located outside the main permeability barrier of the bacteria (see below).

The most important characteristic of MDH is the nature of its prosthetic group which is pyrrolo-quinoline quinone (PQQ) (*Figure 9*). Each monomer of the enzyme contains one molecule of PQQ. Although covalent bonds are not involved in its binding it is attached to the apoenzyme sufficiently strongly that dissociation always leads to destruction of enzyme activity. PQQ is also the prosthetic group of a number of other dehydrogenases, although its binding to these proteins appears to be different. It is probably attached covalently to the apoprotein of primary amine dehydrogenase of some methylotrophs, and in bacterial glucose dehydrogenase it is loosely attached so enabling this enzyme to be used in reconstitution studies.

(ii) *The electron acceptor for methanol dehydrogenase.* All methylotrophic bacteria investigated have been shown to contain at least two types of soluble *c*-type cytochromes. One of these has a high isoelectric point and appears to be similar in structure and function to the typical soluble cytochrome *c* that mediates between the *b*-type cytochromes and cytochrome oxidase in mitochondria and many bacteria. This has been called cytochrome c_H. The second soluble *c*-type cytochrome appears to be unique to methylotrophic bacteria; it has a low isoelectric point and has been called cytochrome c_L. This cytochrome, unlike cytochrome c_H, is the specific acceptor for MDH. Besides having a distinctive low isoelectric point, cytochrome c_L is distinct in having a high molecular weight ($> 17 000$ daltons). The special role proposed for cytochrome c_L in methanol oxidation has been confirmed recently by characterization of a mutant lacking this cytochrome but containing MDH and all other cytochromes; the only characteristic absent in this mutant is its ability to oxidize and grow upon methanol (Nunn and Lidstrom, 1986a,b).

MDH and the two soluble *c*-type cytochromes are located exclusively in the periplasm of those bacteria in which this can be studied. The concentrations of MDH and cytochrome *c* are so high that it has been calculated that there is sufficient of these proteins to form a monomolecular layer over the whole surface of the bacteria. If, as is almost certainly the case, soluble *c*-type cytochromes are always located on the outer surface of bacteria, and if MDH always reacts directly with soluble cytochrome *c*, then an unusual situation occurs in bacteria growing on methane. If the MMO oxidizes methane on the inside of the bacteria then the product, methanol, must diffuse out in order to be oxidized by MDH to formaldehyde which will then have to re-enter the bacteria. Although this must presumably be the case in bacteria having the soluble MMO, it is not necessarily so in those with a particulate enzyme where the reductant NADH might bind on the inside, where it is produced, while the methane and oxygen may react on the outer surface of the membrane.

A preliminary description of an unusual MDH has been published that is also a PQQ-containing enzyme but which may produce NADH as its product (Duine *et al.*, 1984). Such a dehydrogenase would clearly have implications in any bacteria in which growth yields are limited by NADH supply or by the low ATP yield during methanol oxidation by conventional MDH.

(iii) *Regulation of methanol dehydrogenase.* It is often stated that the rate-limiting component of an electron transport chain is determined by the activity of the primary dehydrogenases but this appears not to be the case with MDH in *M. methylotrophus* (Greenwood and Jones, 1986). The rate of electron flow from methanol through MDH will be determined by the concentration of oxidized cytochrome c_L (and hence by the redox potential) but it is probable that some more specific mechanism of regulation is also involved. The variety of metabolic fates of the formaldehyde produced by MDH activity in different methylotrophs suggests that their MDHs, otherwise similar in most other respects, may differ with respect to their regulation, but almost nothing is known about this.

Most MDHs are able to oxidize formaldehyde to formate. This second oxidation step, if allowed to occur, would represent a waste of energy and carbon substrate and so must presumably be prevented. This regulation may be achieved by a modifier protein (M-protein) for MDH. This protein was first shown to account for the oxidation by MDH of 1,2-propanediol, a substrate not oxidized by the pure enzyme. In the presence of M-protein the affinity of MDH for propanediol is increased (Ford *et al.*, 1985). Subsequently it has been shown that the M-protein has a second effect, which may reflect its physiological role, of decreasing the affinity of MDH for formaldehyde, thus preventing its oxidation (Page and Anthony, 1986). Whether or not the formation and location of M-protein itself is regulated and whether or not it has a more complex role than merely preventing formaldehyde oxidation is not yet known.

Electron transport and energy transduction in methylotrophic bacteria

As mentioned above, MDH exerts an important influence on the energetics of methylotrophic bacteria because it interacts with the electron transport chain at the level

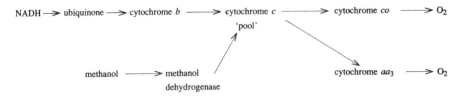

Figure 10. Electron flow in methylotrophs during growth on methanol. The cytochrome c 'pool' probably involves membrane-bround cytochrome c during oxidation of NADH but not of methanol. Cytochrome co is the o-type oxidase. Some methylotrophs have both oxidases; some have only one of them.

of cytochrome c, thus by-passing the low potential part of the electron transport chain containing ubiquinone, iron−sulphur proteins and b-type cytochromes (*Figure 10*). *Table 2* expresses this quantitatively; it can be seen that in all methylotrophic bacteria far less electron flow is by way of an 'NADH oxidase' electron transport chain than in bacteria growing on conventional substrates. In the methylotrophs more than half of the electron transport is by a 'methanol oxidase' system and any special bioenergetic features of this system will be strongly expressed.

Relatively little information has been published on electron transport in methanotrophic bacteria but there is no evidence that it is markedly different from that described in methanol-utilizers. Most information about the special 'methanol oxidase' electron transport chain is available for *Pseudomonas* AM1 (a pink facultative methylotroph) and *M. methylotrophus* (an obligate methanol-utilizer).

In *Pseudomonas* AM1 the predominant oxidase is cytochrome aa_3. This is similar to other oxidases of this type; it oxidizes the small cytochrome c_H but not the cytochrome c_L (Fukumori *et al.*, 1985).

M. methylotrophus is able to produce two oxidases. Cytochrome aa_3 is observed only in bacteria grown under carbon-limited conditions, whereas an o-type oxidase is always present but is induced 10-fold during growth in carbon-excess conditions (Cross and Anthony, 1980; Greenwood and Jones, 1986). This oxidase has been solubilized and completely purified, and shown to consist of equal amounts of b-type and c-type cytochromes. The active oxidase consists of two cytochrome b subunits plus two cytochrome c subunits. The cytochrome c subunit does not correspond to either of the two predominant soluble cytochromes c, nor to the membrane-bound cytochrome c_L also found in the membranes of these bacteria. The purified oxidase oxidizes cytochrome c_H at 50 times the rate of cytochrome c_L. *Figure 11* illustrates the proposed electron transport chain involving the o-type oxidase which, because of the nature of its cytochrome constituents, is now called cytochrome co. From the results quoted above for *Pseudomonas* AM1, it appears likely that a scheme similar to that in *Figure 11* also operates in that organism and in *M. methylotrophus* grown under conditions of carbon limitation, except that cytochrome aa_3 will replace the cytochrome co.

The arrangement of MDH, soluble c-type cytochromes and oxidase illustrated in *Figure 11* constitutes a simple redox arm which is the simplest energy transducing electron transport chain known to be involved in oxidation of an organic compound. This redox arm leads to the release on the outside of the bacteria of two protons per molecule of methanol during electron transfer from MDH to cytochrome c_L; and to consumption on the inside of the bacteria of two protons by the oxidase during reduction of

Table 2. Electron flow from each dehydrogenase expressed as a proportion of total electron transport.

Growth substrate	% of electron flow from each dehydrogenase			
	Methanol	*Formaldehyde*	*Flavoprotein*	*NADH*
Methane				
Serine pathway bacteria (e.g. *Methylosinus trichosporium*)	78–88	(NAD$^+$-linked)	9–12	0–13
RuMP pathway bacteria (e.g. *Methylococcus capsulatus*)	84–92	(NAD$^+$-linked)	0	8–16
Methanol				
Serine pathway bacteria				
(a) HCHO oxidation yields 1 NADH (e.g. *Pseudomonas* AM1)	53–56	35	10	0–3
(b) HCHO oxidation yields 2 NADH	53–64	(NAD$^+$-linked)	9–15	21–40
RuMP pathway bacteria (e.g. *Methylophilus methylotrophus*)	57–70	(NAD$^+$-linked)	0	30–43
RuBP pathway bacteria (e.g. *Paracoccus denitrificans*)	52–62	(NAD$^+$-linked)	0	38–48
Conventional substrates (including formate)	0	0	0–26	74–100

The range of values predicted is for a range of assumed P/O ratios; the lowest is assumed to be one for each oxidation step and the highest P/O ratios are assumed to be 1,2 or 3 for methanol dehydrogenase, flavoprotein and NADH, respectively. Higher P/O ratios lead to a lower proportion of electron transport from NADH dehydrogenase and a higher proportion from other dehydrogenases. During growth on methane only 25–40% of the total oxygen consumed is by way of electron transport and oxidases, the remainder being used in the initial hydroxylation reaction (assumed here to be NADH-linked). In serine pathway bacteria the oxidation of acetyl-CoA to glyoxylate is assumed to involve a flavoprotein. This table is taken from Anthony (1982) with permission of Academic Press.

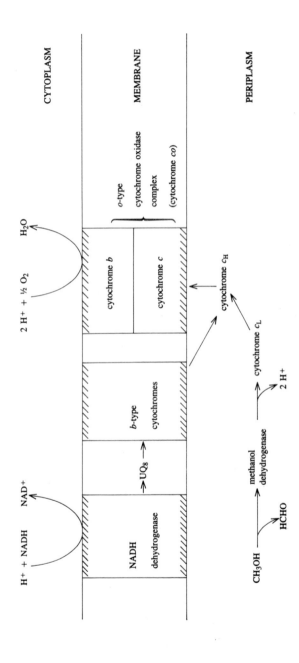

Figure 11. The electron transport chain of *M. methylotrophus* grown in oxygen-limited conditions. A membrane-bound cytochrome *c* probably mediates the oxidation of mid-chain *b*-type cytochromes by cytochrome c_H. There is no evidence for the sequence of *b* and *c* cytochromes in the cytochrome *co* oxidase complex.

oxygen. There is no evidence that either of the two oxidases in *M. methylotrophus* acts as a proton pump. Dawson and Jones (1982) have concluded that this organism is able to sustain a proton motive force (p.m.f., $\Delta\mu H^+$) of up to -165 mV during respiration with methanol as substrate. Either the ΔpH or the membrane potential ($\Delta\psi$) value was shown to be fully competent to drive ATP synthesis. The p.m.f. consisted exclusively of the membrane potential at an external pH of 7.0, which is the same as the internal pH of these bacteria; the phosphorylation potential (ΔGp) at this pH was -48.8 kJ/mol. This is sufficient to yield no more than one molecule of ATP synthesized per molecule of methanol oxidized. This P/O ratio of one (or less) is likely to apply whichever oxidase is operating. The reason for being able to produce two different oxidases in *M. methylotrophus* is not known but it is worth noting that a number of, otherwise similar, obligate methanol-utilizers have only the *o*-type oxidase, cytochrome *co*.

References

Anthony,C. (1978) The prediction of growth yields in methylotrophs. *J. Gen. Microbiol.*, **104**, 91 – 104.

Anthony,C. (1980) Methanol as substrate: theoretical aspects. In *Hydrocarbons in Biotechnology.* Harrison,D.E.F., Higgins,I.J. and Watkinson,R. (eds), Hayden and Son Ltd, London, pp. 35 – 57.

Anthony,C. (1982) *The Biochemistry of Methylotrophs.* Academic Press, London.

Anthony,C. (1986) Bacterial oxidation of methane and methanol. *Adv. Microb. Physiol.*, **27**, 113 – 210.

Beardsmore,A.J., Aperghis,P.N.G. and Quayle,J.R. (1982) Characterisation of the assimilatory and dissimilatory pathways of carbon metabolism during growth of *Methylophilus methylotrophus* on methanol. *J. Gen. Microbiol.*, **128**, 1423 – 1439.

Ben-Bassatt,A. and Goldberg,I. (1980) Purification and properties of glucose 6-phosphate dehydrogenase (NADP/NAD) and 6-phosphogluconate dehydrogenase (NADP/NAD) from methanol-grown *Pseudomonas* C. *Biochim. Biophys. Acta*, **611**, 1 – 10.

Cross,A.R. and Anthony,C. (1980) The electron transport chains of the obligate methylotroph, *Methylophilus methylotrophus*. *Biochem. J.*, **192**, 423 – 429.

Dalton,H. and Leak,D.J. (1985) Methane oxidation by microorganisms. In *Microbial Gas Metabolism.* Poole,R.J.K. and Dow,C.S. (eds), Academic Press, London, pp. 173 – 200.

Dalton,H., Prior,S.D., Leak,D.J. and Stanley,S.J.H. (1984) Regulation and control of methane monooxygenase. In *Microbial Growth on C-1 Compounds.* Crawford,R.L. and Hanson,R.S. (eds), American Society of Microbiology, Washington, USA, pp. 75 – 82.

Dawson,M.J. and Jones,C.W. (1982) The proton motive force and phosphorylation potential developed by whole cells of the methylotrophic bacterium *Methylophilus methylotrophus*. *Arch. Microbiol.*, **133**, 55 – 61.

Douma,A.C., Veenhuis,M., de Koning,M., Evers,M. and Harder,W. (1985) Dihydroxyacetone synthase is localized in the peroxisomal matrix of methanol-grown *Hansenula polymorpha. Arch. Microbiol.*, **143**, 237 – 243.

Duine,J.A., Frank,J. and Berkhout,M.P.J. (1984) NAD -dependent, PQQ-containing methanol dehydrogenase: a bacterial dehydrogenase in a multienzyme complex. *FEBS Lett.*, **168**, 217 – 221.

Ford,S., Page,M.D. and Anthony,C. (1985) The role of a methanol dehydrogenase modifier protein and aldehyde dehydrogenase in the growth of *Pseudomonas* AM1 on 1,2-propanediol. *J. Gen. Microbiol.*, **131**, 2173 – 2182.

Fukumori,Y., Nakayama,K. and Yamanaka,T. (1985) Cytochrome *c* oxidase of *Pseudomonas* AM1: purification and molecular and enzymatic properties. *J. Biochem.*, **98**, 493 – 499.

Fulton,G.L., Nunn,D.N. and Lidstrom,M.E. (1984) Molecular cloning of a malyl-CoA lyase gene from *Pseudomonas* sp. strain AM1, a facultative methylotroph. *J. Bacteriol.*, **160**, 718 – 723.

Goodman,J.M. (1985) Dihydroxyacetone synthase is an abundant constituent of the methanol-induced peroxisome of *Candida boidinii. J. Biol. Chem.*, **260**, 7108 – 7114.

Goodman,J.M., Scott,C.W., Donahue,P.N. and Atherton,J.P. (1984) Alcohol oxidase assembles post-translationally into the peroxisome of *Candida boidinii. J. Biol. Chem.*, **259**, 8485 – 8493.

Green,J. and Dalton,H. (1985) Protein B of soluble methane monooxygenase from *Methylococcus capsulatus* (Bath): a novel regulatory protein of enzyme activity. *J. Biol. Chem.*, **260**, 15795 – 15801.

Greenwood,J.A. and Jones,C.W. (1986) Environmental regulation of the methanol oxidase system of *Methylophilus methylotrophus. J. Gen. Microbiol.*, **132**, 1247 – 1256.

Higgins,I.J., Best,D.J., Hammond,R.C. and Scott,D. (1981) Methane-oxidising microorganisms. *Microbiol. Rev.*, **45**, 556−590.

Lund,J. and Dalton,H. (1985) Further characterisation of the FAD and Fe_2S_2 redox centres of component C, the NADH:acceptor reductase of the soluble methane monooxygenase of *Methylococcus capsulatus* (Bath). *Eur. J. Biochem.*, **147**, 291−296.

Lund,J., Woodland,M.P. and Dalton,H. (1985) Electron transfer reactions in the soluble methane monooxygenase of *Methylococcus capsulatus* (Bath). *Eur. J. Biochem.*, **147**, 297−305.

Leak,D.J., Stanley,S.H. and Dalton,H. (1985) Implications of the nature of methane monooxygenase on carbon assimilation in methanotrophs. In *Microbial Gas Metabolism*. Poole,R.K. and Dow,C.S. (eds), Academic Press, London, pp. 201−208.

Nunn,D.N. and Lidstrom,M.E. (1986a) Isolation and complementation analysis of 10 methanol oxidation classes and identification of the methanol dehydrogenase structural gene of *Methylobacterium* sp. strain AM1. *J. Bacteriol.*, **166**, 581−590.

Nunn,D.N. and Lidstrom,M.E. (1986b) Phenotypic characterisation of 10 methanol oxidation mutant classes in *Methylobacterium* sp. strain AM1. *J. Bacteriol.*, **166**, 591−597.

Page,M.D. and Anthony,C. (1986) Regulation of formaldehyde oxidation by the methanol dehydrogenase modifier proteins of *Methylophilus methylotrophus* and *Pseudomonas* AM1. *J. Gen. Microbiol.*, **132**, 1553−1563.

Prior,S.D. and Dalton,H. (1985) The effect of copper ions on membrane content and methane monooxygenase activity in methanol-grown cells of *Methylococcus capsulatus* (Bath). *J. Gen. Microbiol.*, **131**, 155−163.

Sahm,H. (1977) Metabolism of methanol by yeast. *Adv. Biochem. Eng.*, **6**, 77−103.

Tani,Y., Kato,N. and Yamada,H. (1978) Utilisation of methanol by yeasts. *Adv. Appl. Microbiol.*, **24**, 165−186.

Tatra,P.K. and Goodwin,P.M. (1983) R-plasmid mediated chromosome mobilisation in the facultative methylotroph *Pseudomonas* AM1. *J. Gen. Microbiol.*, **129**, 2629−2632.

Tatra,P.K. and Goodwin,P.M. (1984) R-factor mediated chromosome mobilisation in the facultative methylotroph *Pseudomonas* sp. strain AM1. In *Microbial Growth on C-1 Compounds*. Crawford,R.L. and Hanson,R.S. (eds), American Society for Microbiology, Washington, USA, pp. 224−227.

Tatra,P.K. and Goodwin,P.M. (1985) Mapping of some genes involved in C-1 metabolism in the facultative methylotroph *Methylobacterium* sp. strain AM1 (*Pseudomonas* AM1). *Arch. Microbiol.*, **143**, 169−177.

Veenhuis,M., van Dijken,J.P. and Harder,W. (1983) The significance of peroxisomes in the metabolism of one-carbon compounds in yeasts. *Adv. Microb. Physiol.*, **24**, 1−82.

Woodland,M.P. and Dalton,H. (1984) Purification and properties of component A of the methane monooxygenase from *Methylococcus capsulatus* (Bath). *J. Biol. Chem.*, **259**, 53−59.

Woodland,M.P. and Cammack,R. (1985) Electron spin resonance properties of component A of the soluble methane monooxygenase from *Methylococcus capsulatus* (Bath). In *Microbial Gas Metabolism*. Poole,R.K. and Dow,C.S. (eds), Academic Press, London, pp. 209−213.

Zatman,L.J. (1981) A search for patterns in methylotrophic pathways. In *Microbial Growth on C-1 Compounds*. Dalton,H. (ed.), Heyden and Son, London, pp. 42−54.

CHAPTER 7

Hydrocarbons as feedstocks for biotechnology

J.W.DROZD

Fermentation and Microbiology Division, Shell Research Limited, Sittingbourne Research Centre, Sittingbourne, Kent ME9 8AG, UK

Introduction

In this chapter I deal with a wide range of hydrocarbons and their derivatives (except methanol see chapters 6 and 9, and attempt to illustrate some principles and trends governing their use in biotechnology, not only as carbon feedstocks and starting materials for transformations but also as aids to process operation and product recovery.

A major impetus for the use of hydrocarbons as carbon feedstocks has been their price, this confines their use to the high volume, low price products of biotechnology. Apart from price, there are other advantages and disadvantages to the use of hydrocarbons as feedstocks for these products, which are discussed. The potential for increasing the range of high volume, low price materials produced by biotechnology is examined. Such materials include some novel products of biotechnology, as well as existing petrochemicals. The biological production of propylene oxide is used to demonstrate the advantages and disadvantages of biotechnology for such petrochemical products.

The use of hydrocarbons and their derivatives as starting materials or feedstocks for the production of specialized high value added products such as pharmaceuticals, fine chemicals and intermediates will be discussed. In this area generalizations are difficult to make although it is an area of considerable economic interest and an area where, because of factors such as regio-, and stereospecificity, biotechnology can compete very successfully with organic chemistry.

I shall not deal with agriculturally derived oils and fats, such as tallow, palm oil or soyabean oil in any detail as these are discussed by Stowell (Chapter 8); the main point is that their current price is low, which could make them interesting feedstocks for certain fermentations (Stowell and Bateson, 1983).

I shall discuss the advantages and disadvantages of hydrocarbons and their derivatives in the areas of price, process aspects and purity in turn, and also describe the use and potential use of such products in process technology.

Low cost, high volume biotechnology products
Carbon feedstock price

The price of the main carbon feedstock is of prime importance for those microbial processes where such feedstock costs are an appreciable fraction of the total manufactur-

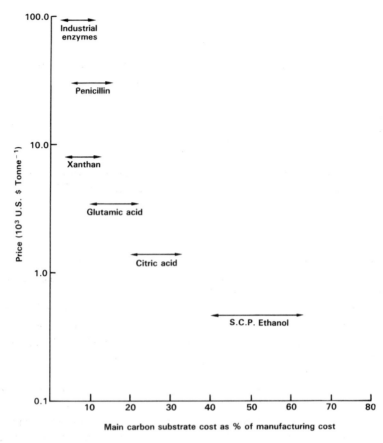

Figure 1. Carbon substrate price as a percentage of manufacturing cost. The data were calculated from Stowell and Bateson, (1983); Dale *et al.* (1984); Cooney, (1979); Van der Beek and Roels, (1984) and were also obtained from a large number of other industrial as well as in house sources. The data for enzyme refers to the weight of enzyme protein.

ing costs (*Figure 1*). For these processes the microbial oxidation of the carbon source supplies energy as well as structural carbon. On a weight basis hydrocarbon feedstocks such as *n*-alkanes have approximately twice the carbon and three times the energy content of sugar. As will be discussed later, nearly all processes where the carbon feedstock price is important are for the manufacture of high volume, low cost products of biotechnology (*Figure 2*).

In the early 1960s and 1970s there was a price incentive to consider using oil or natural gas-derived hydrocarbons as feedstocks for several processes (e.g. see Dimmling, 1976). The oil price was very low and world sugar prices were high (*Figure 3a,b*) although it is unclear as to how much the latter influenced events. The oil price rise in the 1970s dramatically altered the picture and oil-derived hydrocarbons became unattractive as feedstocks (Hamer, 1985). Even though the oil price has recently fallen dramatically it coincides with a low world sugar price so that the differential is still not as great as it was during the early 1960s and 1970s (*Figure 3b*). However if the

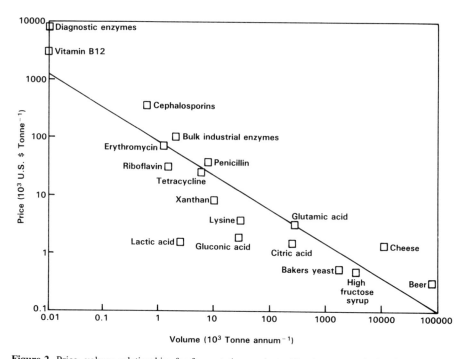

Figure 2. Price, volume relationship, for fermentation products. The data were calculated from Dunnill, (1981); King, (1982); Cooney, (1979); Van der Beek and Roels (1984); Dale *et al.* (1984); Dimmling and Nesemann (1984) and were also obtained from a large number of other industrial as well as in house sources. The data for enzymes refers to the weight of enzyme protein.

oil price were to fall further and the sugar price were to rise then hydrocarbons would again be of interest from a price point of view as feedstocks for some fermentations. Price fluctuations in all feedstocks means that feedstock flexibility for a process would be a distinct advantage and allow the manufacturer to use the most attractive feedstock available at the time. A note of caution is required in looking at *Figure 3a,b*. Firstly, world sugar prices do not reflect the price paid by many consumers (Brown, 1983; Anon, 1985); sugar prices are subject to tariffs, quotas, trade restrictions and agreements. For example the current EEC sugar price is well above the world price (*Table 1*) due to the application of the Common Agricultural Policy (CAP). Secondly, unlike sugar prices, molasses prices in the EEC are more directly linked to world prices and world molasses prices do not show the marked price fluctuations of world sugar prices (*Figure 3a*; Brown, 1983). Thirdly, world sugar prices do not include the large amounts of starch-derived glucose syrups and dextrose available as feedstocks (Dimmling and Nesemann, 1985). Like sugar these are variable in price and in certain areas, for example the EEC, are subject to complex pricing policies (Denyer, 1983). At present glucose syrups and dextrose are competitively priced against sucrose within the EEC (*Table 1*). Fourthly, the liquid *n*-alkane prices shown do not represent lower long-term contract prices but they do indicate relative trends and are closely linked to the oil price.

The technical impetus for the use of hydrocarbons in fermentation processes came from the work of Champagnat *et al.* (1963) whose initial work on a microbial route

121

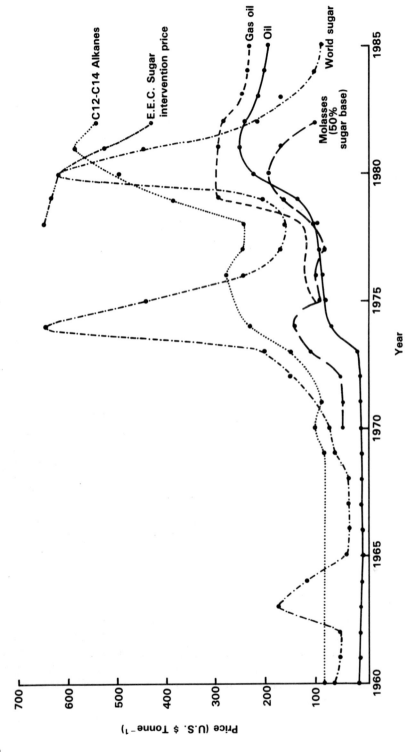

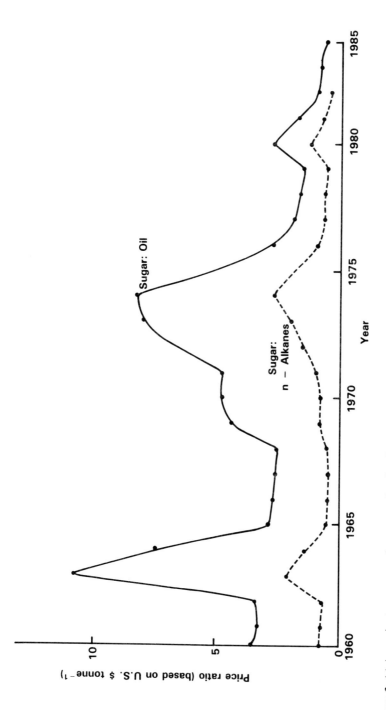

Figure 3. (a) Approximate current yearly average prices for oil, sugar and related products. Western European crude oil prices from the Chemicals Economics Handbook (1985), SRI International and from Energy Statistics Special Edition (1974), EEC, Luxembourg. Gas oil prices from Platt's published European yearly mean prices, FOB, Rotterdam and the Financial Times. U.S. List prices of C12 to C14 *n*-alkane from Chemicals Economics Handbook (1983), SRI International. European prices are shown in Dimmling and Neseman (1984). World Sugar prices were obtained from the World Bank, Commodity Trade and Price Trends (1981), The Economist, The Financial Times, and Brown (1983). Molasses prices were from Brown, (1983); Dale *et al.* (1984); and industrial sources. **(b)** Ratio of current yearly average sugar to crude oil and sugar to *n*-paraffins (C12 to C14 *n*-alkanes) prices from **(a)**.

123

Table 1. Approximate relative prices as of early April 1986.

	US$ per tonne
Crude oil[a]	65 – 88
Gas oil	120 – 130
n-Alkanes	160 – 220
World sugar	180 – 220
EEC sugar[b]	490 – 560
EEC dextrose[c]	450 – 520
EEC molasses[d]	250 – 300
Palm oil	220 – 280

[a]$9 – 11 per Barrel.
[b]As for fermentation use.
[c]Starch derived, as for fermentation use.
[d]Per tonne carbohydrate.
Exchange rate of £1.0 = US$1.47 used.

Data from The Financial Times and sources within the industry. The oil price is based on $ 9 – 11 a barrel. The *n*-alkanes price is calculated from the ratio in *Figure 3a*.

Table 2. Microbial fermentation products which can be produced from hydrocarbons.

Feedstock	Liquid n-alkanes	Methane/natural gas
Product	SCP	SCP
	Organic acids especially citric acid	
	Amino acids especially glutamic acid	
	Carbohydrates and related	
	Nucleic acid and related	
	Vitamins, co-enzymes etc.	
	Antibiotics	

List in approximate order of degree of R and D done on the process.

From Fukui and Tanaka, (1980); Miall, (1980); Ratledge (1984).

for de-waxing gas oil lead to the development of gas oil-based, and subsequently to liquid *n*-alkane-based single cell protein (SCP) processes (Shennan and Levi, 1974). Many other hydrocarbon-based processes (*Table 2*) were investigated, especially in Japan (Fukui and Tanaka, 1980) but also in Europe (Miall, 1980). The economic incentive for the work was the substrate price and availability, and the processes thus developed were for products where feedstock costs represented a major factor in production costs (*Figure 1*). Products in this category are of large production volume and low value (Dunnill, 1981; King, 1982) such as SCP, glutamic and citric acids (*Figure 2*). However, even for antibiotics such as penicillin, substrate price is also a significant part of manufacturing costs (Cooney, 1979; Van der Beek and Roels, 1984) because of the low penicillin yield on the substrate. A similar situation occurs in the production of some industrial enzymes. For high value enzymes or genetically engineered protein products feedstock costs represent only a very small proportion of production costs; downsteam work-up and purification costs dominate manufacturing costs (Datar, 1986).

Figure 2 indicates that many of the high volume, low price products of biotechnology are complex mixtures, for example beer, cheese, which are destined for the human food and beverages market (Dunnill, 1981). These products are manufactured from complex agriculturally derived feedstocks which help to impart their unique organoleptic properties, and so they could not be made from hydrocarbons. SCP is a complex mixture which can be made from hydrocarbons but it has not attained a large sales volume because of escalating feedstock prices in the middle to late 1970s coupled with a fall in the price of competitive animal feeds, such as soyabean meal. The latter makes the economics of SCP production in Western Europe unattractive at the moment; even though oil prices have fallen soyabean meal is only about $200 a tonne. Perhaps the production of the biological polymer poly-β-hydroxybutyrate (PHB) (Collins, Chapter 9) will develop into a high volume industrial sector product.

There are a limited number of simple bulk chemical products of biotechnology (*Figure 2*; Pape, 1976), for example citric and glutamic acids, and these are also mainly used in the human food and beverages industries where there may be acceptability problems with products derived from perceived 'non-natural' hydrocarbon substrates. The only new high volume product of biotechnology is high fructose syrup which is again for the food and drinks market and uses glucose as feedstock. An example of a simple bulk product not destined for this market is fermentation ethanol for use in automobile fuels, but for technical and economic reasons this would not be made from hydrocarbons. Thus there are at the moment only a limited number of high-volume products for which hydrocarbons would be a suitable substrate if the price was right.

The current low price of natural oils and in particular of palm oil (*Table 1*) which is currently undercutting soyabean oil, could make these feedstocks interesting substrates and competitive with *n*-alkanes. Also, there may be fewer acceptability problems as they are of 'natural' origin.

Natural gas could be a cheap substrate in certain locations (Hamer, 1985) where it is currently flared in large amounts (e.g. in the Middle East), but there has been little work on the possible use of natural gas as a feedstock for processes other than SCP production. The conversion of natural gas to methanol and the use of methanol as a feedstock is technically more attractive than the direct use of natural gas.

Future substrate pricing is difficult to predict. For agriculturally derived feedstocks there are complex climatic, geographical and political factors. Within the EEC there are likely changes in the CAP which should bring sugar prices to industry closer to world prices. On a wider front the development of a cheap cellulose hydrolysis process could strongly influence sugar prices. The future oil price is similarly difficult to predict but in the near future the price is unlikely to reach the high levels that we have seen over the last decade. In the longer term prices will rise as known reserves become depleted and more expensive sources exploited. It is the timing of these events which is particularly difficult to predict.

Availability and supply

In the past, the production of agriculturally-derived feedstocks has tended to fluctuate due to factors such as climatic variations and the seasonal nature of crop growth and the harvesting cycle. During the SCP development era in the 1970s, the view developed that hydrocarbons and hydrocarbon derivatives were concentrated feedstocks whose supply was not subject to such fluctuations. The West European n-alkane production capacity in the mid-1970s was over 10^6 tonnes, with potential for further expansion (Dimmling, 1976). While this view may have been sound at that time, the present situation is less clear-cut; in many areas, the EEC for example, sugar and starch are likely to be in surplus in the immediate future (Dimmling and Nesemann, 1985), while, if oil becomes traded as a commodity product, its supply position could fluctuate, with small imbalances between production and demand having major effects on supply and price. Thus, the present advantages of hydrocarbons in this respect may not be maintained.

Purity

Purity depends on the feedstock in question. By definition the more refined feedstocks are purer whether they are oil, gas or agriculturally based. However, agriculturally derived food and beverage products will generally be more acceptable to the general public as they represent products of 'natural' origin.

Gas oil, crude n-alkane cuts and natural gas contain several components. The presence of some components can strongly influence process economics. The British Petroleum SCP from n-alkanes project in Sardinia was terminated in part because some lobbies claimed that toxic and carcinogenic residues from n-paraffins were carried through the food chain. These claims were ill-founded and were refuted by the scientific evidence (Hyde, 1978; Hamer, 1981; Shennan and Levi, 1974). Nevertheless there is a problem of acceptability with oil-based products destined for the food and beverages market. In this context natural oils such as palm oil may give less problems. For SCP production the problem is accentuated by the tendency of hydrophobic cell components to partition out these impurities. Similar considerations apply in other cases, for example for citric and glutamic acid production a high purity paraffin has to be used, especially when the product is destined for the food industry (L.M.Miall, personal communication). For SCP production from natural gas the natural gas has to be freed of commonly occurring higher alkanes such as ethane, if the use of a mixed microbial culture is to be avoided. If higher alkanes are present they are oxidized to corresponding alcohols, aldehydes and acids which will accumulate in a pure culture and inhibit the growth of the methane-utilizing bacteria (Drozd and McCarthy, 1981; Linton and Drozd, 1982).

Thus the statement that purity is an advantage with hydrocarbons should be viewed with circumspection and depends on the feedstock in question. For oil-based feedstocks the purity and price increases with the degree of refining, going from crude oil to gas oils and kerosenes (both middle distillates) to purified n-alkane cuts. As far as hydrocarbon derivatives are concerned both synthetic ethanol and methanol represent pure feedstocks of interest. Purity is of importance if a large percentage of feedstock is not utilized as this effectively influences yield, downstream processing, site location and effluent treatment. For example with gas oil only a maximum of 20% may be readily utilized in a microbial process and hence the fermentation plant should be located close to a refinery which can process the unused feedstock. Cheap agriculturally-derived feedstocks such as molasses are also very impure (Dimmling and Nesemann, 1985) and contain a lot of undefined organic material as well as herbicides, pesticides and residues of chemicals used in processing. However the residues do not seem to cause any animal or human toxicological problems. They can, however, prove deleterious to some fermentations and testing of each molasses batch is often necessary. Molasses is also very variable and causes a considerable effluent treatment problem because of the large amount of non-fermentable organic material present. Of the $70-80\%$ (w/w) total solids present in molasses only about 50% are fermentable sugars (Dimmling and Nesemann, 1985). The use of purified hydrocarbons could help reduce the considerable effluent treatment problems associated with the fermentation industry (Dunnill, 1981) because residues are minimal and because an immiscible liquid can be readily separated out downstream. In this context the purer hydrocarbons are competing against the purified carbohydrate feedstocks, that is against glucose syrups and crystalline sucrose rather than against molasses. Gas oil or some unprocessed natural gases are more akin to molasses in their state of purity. For process development and operation defined purified feedstocks are highly advantageous as there is less variability, less process effluent, easier downstream product purification and the fermentation process can be better defined allowing rational rather than empirical improvements to be made. However for some processes, for example for antibiotic fermentations, complex agriculturally-derived feedstocks such as corn steep liquor provide a host of defined and undefined growth factors which stimulate the process. It would be difficult to develop a defined medium for such a process.

Handling and storage

Possible problems of handling and storage will depend on the feedstock in question. Flammability and explosion hazards are obviously more of a problem with hydrocarbons, such as natural gas and air mixtures, than with sugars but do not generally represent a mjaor difficulty, especially with liquid *n*-alkanes or methanol. Safety measures required for storing and handling hydrocarbons may however give rise to extra costs. Liquid *n*-alkanes and methanol can be stored close to the site of use while natural gas could be continuously supplied by pipeline. The question of substrate sterilization must also be closely considered, especially for water immiscible liquids.

Microbial contamination

Far fewer organisms are capable of growing on hydrocarbons than on sugar and they are therefore much less susceptible to microbial contamination. This gives hydrocarbons a potential advantage over sugars as substrates in that their use should make it easier to operate a continuous or semi-continuous (fill and draw) culture process without risk of microbial contamination. This is of particular importance with large volume, low price products where manufacturing costs are most sensitive to process variables. For example, from this point of view, it is probably easier to operate a continuous SCP process with natural gas, methanol or *n*-paraffins as substrate than with glucose or sucrose. However real comparisons are difficult to make as little published data are available from pilot or manufacturing scale continuous or semi-continuous operations. Also, carbon-limitation is used in SCP production so there is little excess carbon substrate available for growth of contaminants (Linton and Buckee, 1977). However, in the manufacture of products such as amino acids, the amino acids would themselves be readily assimilable substrates. For products such as citric acid, carbon excess condi-

127

tions would be used and contaminants could oxidize the citrate with a consequent loss in yield. For products intended for human food use, care would have to be taken that contaminants were non-pathogenic and did not produce toxins which were associated with the product. Nevertheless despite these drawbacks there could be an incentive to develop some continuous processes with hydrocarbons as feedstocks provided substrate cost was attractive and the process heat output and oxygen demand did not impose an undue economic penalty.

Microbial toxicity

The inhibition of growth of methane-utilizing bacteria by higher alkanes in natural gas, for example ethane and propane was discussed under 'Purity'. Microbial growth on liquid alkanes may be inhibited by certain hydrocarbons, especially alkanes in the C_5-C_9 range (Gill and Ratledge, 1972). This is related to the aqueous solubility of these hydrocarbons. In practice it is unlikely to be a problem because the potentially toxic compounds will be kept at a low concentration either by dilution in the less toxic higher alkanes or by slow metabolism. Freely soluble purified hydrocarbon derivatives such as methanol are toxic if at too high a concentration (Brooks and Meers, 1973). This means that careful process control is required to maintain the methanol concentration within a suitable range. This may be critical for product formation where carbon excess conditions are required.

Physical nature of substrate

If a gaseous substrate, for example methane, is used then there are:
(i)   explosion hazards in the gas phase to be avoided;
(ii)   mass-transfer requirements to get the substrate into solution, and
(iii)   a low conversion per pass in the reactor.

A very high conversion cannot be attained at a single pass while maintaining a high rate of gas mass-transfer and a high process productivity. The off-gas could be used (after enrichment) as a source of energy, or gas phase recycle could be employed. These points are discussed by Hamer (1979) and Hamer and Hamdan (1979).

The use of volatile liquid hydrocarbon feedstocks, such as *n*-pentane or *n*-hexane from liquified petroleum gas heavy ends will introduce major physical and biological problems (Hamer *et al.*, 1980), although for specific locations they may be readily available. Either advanced, novel process technology would have to be used as was envisaged for SCP production from natural gas or else the substrate would have to be chemically converted into a more attractive feedstock such as methanol.

Liquid hydrocarbons, for example $C_{10}-C_{18}$ *n*-alkanes or similar are water immiscible and require to be dispersed in the aqueous phase. There is a large amount of literature on the mixing and disperson of, and microbial growth kinetics with water-immiscible liquid alkanes (e.g. Humphrey and Erickson, 1972; Seipenbusch and Blenke, 1980). What is often unclear is the relationship between volumetric productivity and residual substrate concentration. For a freely soluble substrate, such as sugar or methanol, the productivity will be independent of substrate concentrations down to sugar concentrations of a few mM or less because of the low affinity of the microorganisms for the sugar. For a water-immiscible substrate when the batch culture productivity is initially limited by the intrinsic activity of the microorganisms, there may be a decrease in productivity with time if the productivity becomes limited by substrate interfacial area as this decreases during the fermentation. To maximize yield it is important to utilize all the substrate without sacrificing production or to re-cycle any unused substrate.

An advantage with a water-immiscible carbon substrate is that high product concentrations can be obtained in the reactor, for example over 10% (w/w) for citric acid, without having to add further carbon substrate batchwise, or fed-batchwise as is the case with sugars (too high initial sugar concentrations can inhibit many microbial processes). There is however a penalty in that the addition of large amounts of water-immiscible hydrocarbon will decrease the effective working volume of the reactor.

With a water-immiscible substrate downstream purification steps may be simplified as simple physical separation steps can be used to remove any unused substrate although the product may require special treatment to remove any residual substrate and unacceptable residues. Hydrocarbon derivatives such as alcohols, for example methanol, are of interest because of their complete water miscibility which means there are no substrate dispersion problems.

Microbial technology

This relates to the state of development of existing microbial processes with hydrocarbons as substrate and has largely been discussed in other sections. Most development work has been done with liquid *n*-alkanes (*Table 2*) for which a wide variety of processes and products such as SCP, organic acids, amino acids and even antibiotics have been investigated (Miall, 1980; Fukui and Tanaka, 1980). Far less work has been done on other hydrocarbons, and with natural gas and methane only SCP production has been extensively studied (Linton and Drozd, 1982). Most of the processes researched into are bacterial or yeast based. Compared with fungal processes these have the advantage of a low broth viscosity which minimizes mixing and oxygen transfer problems but the disadvantage is that the downstream separation of biomass is more difficult than with a fungal process.

Heat output, oxygen demand and yield

The reduced nature of hydrocarbons compared with sugars means that for oxidized products such as organic acids and amino acids more oxygen will be required and more heat evolved for a hydrocarbon-based process than for an equivalent sugar-based process. For a reduced compound, for example a microbial lipid, the situation is reversed. *Table 3* illustrates these points and shows the relevant data for the production of a glucose homopolysaccharide, citric acid (glutamic acid gives similar results) and palmitic acid (representative of a microbial lipid) from glucose, methanol, methane and a C_{16} alkane. The data are calculated from the known biochemical pathways of the organisms concerned (Dr J.D.Linton, personal communication). The production of polysaccharides from methanol, methane or liquid *n*-alkanes is unattractive because of the high oxygen demand and heat output, despite the high weight yield on the C_{16} alkane. The highly viscous nature of the broth hinders oxygen and heat mass transfer which can already be a problem with glucose as substrate. Citric acid and glutamic acid production from

Table 3. Substrate yield, oxygen requirement and heat production for microbial product formation from glucose, methanol, methane and a C_{16} n-alkane.

Metabolic product	Carbon source	Yield from carbon source (g/g substrate)	YO_2 (g/g)	Heat of fermentation (ΔH, kcal/mol. product)
Exopolysaccharide	glucose	0.85	16.22	35
(homo glucose,	CH_3OH	0.82	1.57	376
per glucose)	CH_4	0.84	0.28	1844
	C_{16}alkane	1.43	0.88	606
Citric acid	glucose	1.06	4.0	198
	CH_3OH	1.0	1.33	550
	CH_4	1.25	0.66	1414
	C_{16}alkane	2.25	1.39	487
Palmitic acid	glucose	0.34	3.8	402
	CH_3OH	0.33	0.61	1701
	CH_4	0.32	0.22	7251
	C_{16}alkane	1.13	8.0	161

The values are for product formation from non-growing cells and were calculated from the known biochemical pathways of the organisms concerned.

a C_{16} alkane or methanol are not too unfavourable compared with glucose as substrate. For palmitic acid production, a C_{16} alkane is the preferred substrate as the feedstock is only partially oxidized. For these reasons products like biosurfactants are also attractive targets to be made from liquid n-alkanes.

The data in *Table 3* do not include culture growth requirements. For these the oxygen demand and heat output per unit of biomass will increase with the degree of substrate reduction. The rates of oxygen and heat transfer required will depend on the process productivity and design, for example whether product is formed during or after culture growth.

Productivity

For high volume, low cost products it is important to maximize productivity as this influences capital costs. It has been pointed out already that the use of hydrocarbons may give a process less susceptible to microbial contamination and hence enable productivity to be increased by use of fill and draw or fully continuous operation. Unfortunately, data on the comparative specific [per g product (g biomass) per hour] or volumetric [per g product per hour] productivities of sugar- and hydrocarbon-based processes for products such as citric and glutamic acids are scarce.

Suzuki et al. (1971) produced 82 g/l of glutamic acid in 48 h from n-paraffins. Akiyama et al. (1973) produced 80 g/l citric acid in 48 h from n-paraffins. Both these productivities are promising.

Higher value-added products

Hydrocarbons and their derivatives will show further potential as feedstocks and starting materials for the production of higher value added products in the pharmaceutical, agricultural chemical, food, fine and speciality chemicals areas.

Broadly there are two related and overlapping areas:

(i)   Where biotechnology is used to transform hydrocarbons or their derivatives into the desired product. In this area biotechnology enables complex stereo-, or regio-specific transformations to be undertaken by enzymes or cells which may not necessarily be grown on, or be able to grow on the substrate in question. A wide range of reactions are possible (e.g. Kieslich, 1976; Tramper *et al.*, 1985), some examples being the microbial conversion of naphthalene to salicyclic acid (Miall, 1980), the microbial transformation of steroids and the resolution of the esters of menthyl acetate (Brookes *et al.*, 1986). Recent developments in the understanding of oxygenase reactions (Dalton, 1980) have added to the possibilities of carrying out specific oxidations and epoxidations of compounds by methane-oxidizing bacteria. The use of lipases for inter-esterification of natural fats, for example to produce cocoa butter-type triacylglycerols from cheaper natural oils and fats is another example (Ratledge, 1984). Very many other examples can be found in the literature.

(ii)   Where hydrocarbons are the necessary fermentation feedstocks for the production of certain speciality lipids, long chain fatty acids, waxes, biosurfactants, etc. (Rehm and Reiff, 1981; Ratledge, 1980, 1983, 1984).

The above examples include products of high and intermediate value. For intermediate value compounds the viability of the biotechnological route will depend on the economics of any competing chemical route and for performance chemicals, such as biosurfactants, the relative performance of the product.

In general, for the production of high value compounds, productivity and hence cooling and oxygen transfer, are not nearly so critical as they are for high volume, low value products. For many specialized transformations the cost of the starting material may be significant (Stowell and Bateson, 1983) or product recovery and purification may be the major process costs. A growing number of specialized biotransformations will be undertaken by commercial organizations as part of complex in house chemical synthesis programmes.

Biotechnology and bulk petrochemicals: propylene oxide production as an example

It has been debated whether biotechnology can compete with existing processes for the production of bulk petrochemicals (Drozd, 1980).

The possibility of a biotechnological route for propylene oxide formation has attracted recent interest (e.g. Habets-Crützen *et al.*, 1984; Wingard *et al.*, 1985; Neidleman *et al.*, 1981). As an example of a large volume, cheap product (by conventional technology approximately £800 a tonne when produced from propylene at £400 a tonne) it is interesting to look at the advantages and disadvantages of the biotechnological routes. The biotechnological routes show similarities to the chemical routes (*Figure 4*) but compared with the chemical routes the biotechnological routes suffer major disadvantages in the following.

(i)   Low volumetric reaction rates for biological routes 1−3 (*Figure 4*) which involve alkane-, or alkene-oxidizing bacteria (Drozd and Bailey, unpublished). For example, the maximum *in situ* rate of ethylene oxidation by ethylene-oxidizing

Chemical Routes:

1. Chlorohydrin route =

$$Ca(OH_2) + 2Cl_2 + H_2O + 2 \text{ Propylene} \longrightarrow CaCl_2 + 2HCl + 2 \text{ Propylene oxide}$$

2. Hydroperoxide Co-product process:

Ethylbenzene + O_2 + Propylene $\longrightarrow$ Styrene + Propylene oxide + H_2O

Some Potential Biotechnological routes[1]

1. Methane-oxidising bacteria:

$$2CH_4 + 2 \text{ Propylene} + 5O_2 \longrightarrow 2 \text{ Propylene oxide} + 2 \text{ } CO_2 + 4H_2O$$
$$2CH_3OH + 4\text{Propylene} + 5O_2 \longrightarrow 4 \text{ Propylene oxide} + 2CO_2 + 4H_2O$$
$$2C_2H_5OH + 2 \text{ Propylene} + 3O_2 \longrightarrow 2 \text{ Propylene oxide} + 2 \text{ Acetate} + 2H_2O$$

2. Propylene-oxidising bacteria:

$$8 \text{ Propylene} + 8O_2 \longrightarrow 7 \text{ Propylene oxide} + 3CO_2 + 3H_2O$$

3. Ethylene-oxidising bacteria:

Ethylene + 4 Propylene + $5O_2 \longrightarrow$ 4 Propylene oxide + $2CO_2$ + $2H_2O$

4. "CETUS" route:

Propylene + O_2 + Glucose $\longrightarrow$ Propylene oxide + Gluconic acid
Propylene + O_2 + $CH_3OH \longrightarrow$ Propylene oxide + H.CHO + H_2O

[1] Assuming 100% coupling of reductant generated by substrate oxidation
to propylene oxide production, and no microbial growth.

Figure 4. Routes for propylene oxide formation. The stoicheiometries of the biotechnological routes are theoretical assuming a 100% coupling of reductant generated by substrate oxidation to propylene oxide formation.

bacteria growing at μmax of 0.04/h is approximately 1.8 mmol g dry wt/h which with a biomass hold-up of 60 g/l gives 108 mmol l/h. If these bacteria would support 80 mmol propylene oxide formation per litre per hour then a reactor volume of 3000 m^3 would be required to produce 100 000 tons of propylene oxide per annum assuming a stream time of 300 days. For methane-oxidizing bacteria a higher reaction rate is possible because of a 10-fold higher μmax but in all these cases the maximum oxidation rate only occurs in growing cultures. In practice, because of the toxic nature of the product and the competitive nature of the reaction on growth only non-, or slowly-growing, cultures will be used in which the reaction rates are lower still and hence yet larger reactor sizes will be required.

(ii) The reaction product propylene oxide, rapidly and irreversibly destroys enzymatic activity at low concentrations (Habets-Crützen and de Bont, 1985). For example (Drozd and Bailey, unpublished) even when continuously stripped from the reactor 1 g dry weight of ethylene utilizing bacteria catalysed the production of a maximum of only 0.1 g of propylene oxide. This can be compared with the production of 50 g of ethanol from glucose by 1 g of non-growing yeast cells (Drozd and Godley, unpublished). Methane-oxidizing bacteria are even more sensitive to propylene oxide than ethylene-oxidizing bacteria and thus ways would have to be developed to minimize product inhibition.

Table 4. Theoretical process parameters for biological[c] propylene oxide formation

Propylene oxide production rate [mmol/l/h]	Substrates	Products	Heat evolved[b] [Kcal l/h]	Required O_2[b] transfer rate [mmol/l/h]	Reactor volume[a] (m³)
15	CH_4, O_2, C_3H_6	CO_2, C_3H_6O	3.6	37.5	16×10^3
1500			362	3750	160
15	C_2H_4, O_2, C_3H_6	CO_2, C_3H_6O	1.7	18.8	16×10^3
1500			171	1880	160
15	O_2, C_3H_6	CO_2, C_3H_6O	1.5	17.1	16×10^3
1500			150	1714	160
15	$C_6H_{12}O_6$, O_2	C_3H_6O, $C_6H_{12}O_7$	1.5	15	16×10^3
1500	C_3H_6		150	1500	160

[a]For 100 000 tonnes per annum with stream time of 300 days.
[b]For comparison a high productivity [5 g/l/h] methanol based SCP process requires an O_2 transfer rate of approximately 300 mmol/l/h and evolves 36 Kcal/l/h heat. With the same conditions as above for 100 000 tonnes per annum a reactor volume of 2.8 × 10³ m³ is required.
[c]For comparison a 100 000 tonnes propylene oxide per annum chemical process with styrene co-product has a reactor volume of approximately 200 m³.
The data were calculated using the equations in *Figure 4*. A biomass hold up of 60 g/l was assumed with the attainable specific product formation rate of 0.25 mmol/(g dry wt bacteria)/h or an increased rate of 25 mmol/(g dry wt bacteria)/h. Air is the gaseous phase and the dissolved oxygen concentration is below 0.5% of air saturation.

(iii) A very high rate of oxygen transfer is required (*Table 4*). This is difficult to attain economically in an aqueous environment in which oxygen has a limited solubility.

(iv) A correspondingly high rate of heat removal is necessary (*Table 4*). In this respect operation at temperatures of below 50°C is disadvantageous and is a major drawback to the biological route.

If cell immobilisation is considered as a way of retaining biomass in the reactor then problems of product toxicity and of heat and oxygen mass-transfer will be heightened (Drozd, 1980).

Process evaluations emphasize that to approach economic viability the reaction rate has to be increased and problems of product toxicity decreased while there has to be a market for any biomass co-product; the value of this biomass has an important influence on process economics. The toxicological problems surrounding the use of *n*-alkanes as a substrate for SCP production suggest that similar problems might be associated with the use of SCP which had been exposed to propylene oxide. Similar conclusions were arrived at by Exxon (Hou, 1984).

The Cetus route (equation 4, *Figure 4*) is interesting and is similar to the chemical chlorohydrin route. It utilizes chloroperoxidase and epoxidase enzymes. Hydrogen peroxide is required and can be generated biologically, for example in oxidizing glucose to gluconic acid via glucose-1-oxidase or methanol to formaldehyde via methanol oxidase. In theory much higher volumetric reaction rates than the other biological routes are attainable but problems of reactant and product toxicity remain and the problems of heat and oxygen mass transfer are heightened.

What might make the alkene-oxidizing bacteria mediated route interesting is its ability to produce an enrichment in one of the optical isomers of propylene oxide (Habets -

Crützen *et al.*, 1985). Such a product would have a higher value than the standard propylene oxide product and could form the starting point of some interesting stereochemical syntheses.

The case of microbial propylene oxide production clearly illustrates many of the disadvantages of biotechnology as an alternative to current chemical process technology for the production of simple high volume, low value petrochemical products. The disadvantages of biotechnology in producing simple bulk petrochemicals was also demonstrated by Pape (1976) in his comparison of the biological and chemical routes for acetic acid production. In general biotechnological routes cannot compete with existing routes for the manufacture of such products.

Process technology

Hydrocarbons and their derivatives do have a unique role in biotechnology not as feedstocks, but as aids to processes. There are several examples.

(i)   Liquid *n*-alkanes or other suitable non water-miscible compounds can be added to fermentations to minimize inhibition caused by hydrophobic products. Such multi-phase fermentations can be used to partition out a toxic product provided that other components do not partition out as well. This is most critical for higher volume, lower price products where high productivity is of importance in reducing manufacturing costs. Several examples are given in the review of Brink and Tramper (1985). However for reasons already discussed many of these processes are unlikely to be economic even if product inhibition is minimized by the use of multiphase fermentations. A related example, although it is not product inhibition in the strict sense, is in polysaccharide fermentations where the final product concentration is often limited by the rate of transfer of oxygen into and of heat from the highly viscous fermentation broths. In practice this occurs at 1.5−3% polysaccharide. The use of a water in oil, for example paraffin emulsion has the advantage that there is no viscosity development in the bulk oil phase (Engelskirchen *et al.*, 1985; Voelskow *et al.*, 1985). There are the following disadvantages.

(a)   That the working reactor volume is effectively reduced by at least 50% and hence there is a liquid phase productivity penalty;

(b)   It is difficult to control the pH in the aueous phase droplets; extremes of pH can cause a loss in productivity and a loss in product performance. Conventional pH control systems do not give a rapid and even control of pH in the aqueous phase.

(c)   It is difficult to add extra glucose to the aqueous phase. The whole idea is that high polymer concentrations (>3%) can easily be reached, which requires more than 6% sugar. As most polysaccharide-producing bacteria are adversely affected by more than 3% sugar it becomes necessary to add the extra glucose during the fermentation and to ensure that it is rapidly and evenly mixed in the aqueous phase droplets. Thus both pH control and sugar addition are problems to be addressed.

(ii)   The recent development work by Kula and her associates (Kroner *et al.*, 1984) indicate that the use of multiphase systems, such as a polyethylene−glycol phase and a saline aqueous phase may give economically attractive routes for the large-scale enzyme recovery and enrichment. To improve the economics the overall chemicals consumption should be minimized and the starting cell concentration maximized. Kroner *et al.* (1984) report that two such processes are operating industrially.

Summary

At present, except in certain geographical situations, the price of hydrocarbons does not make them economically attractive feedstocks for the production of established biotechnology products where feedstock costs are an appreciable fraction of manufacturing costs. Such products are mainly destined for the food and beverages markets (*Figures 1 and 2*) where agriculturally-derived feedstocks, including natural oils, may be more acceptable. Although the price of oil is falling, this also coincides with a low world sugar price (*Figure 3*) and the differential would have to increase further to make hydrocarbons economically attractive. In certain geographical locations where sugar prices are high, for example in the EEC, the situation is more favourable for hydrocarbons (*Table 1*). If hydrocarbons do become attractive on a substrate−cost basis, then to be economically attractive any increased capital (if reactor productivity is decreased) and operating costs (e.g. oxygen transfer, heat removal) have to be taken into consideration.

Although molasses is used for many large-volume fermentation processes, it is in many ways unsatisfactory and higher purity feedstocks such as crystalline sucrose or glucose syrups would be preferred if the price was right. In a similar way unpurified hydrocarbons, such as gas oil would be unsuitable and a refined product such as a liquid *n*-alkane cut or methanol may be preferred.

In general, because the ability of biotechnology to compete with existing chemical routes for the production of simple, high volume petrochemicals is poor, there will be few new biotechnological processes developed in this area where hydrocarbons or their derivatives are the necessary process feedstocks. Also, despite extensive work few other new large volume products of biotechnology have appeared. We can thus expect relatively few new biotechnological processes to be developed where substrate price is a large percentage of manufacturing costs and hence few potential new uses of relatively large amounts of hydrocarbons as feedstocks in biotechnology.

Hydrocarbons and their derivatives do have a role in microbial process technology, for example in multi-phase enzyme extractions, in multi-phase fermentations and in product recovery processes.

Hydrocarbons and their derivatives have further potential as starting materials or feedstocks in biotechnology for the production of specific higher value products such as intermediates, pharmaceuticals, fine chemicals, speciality chemicals and agricultural chemicals. Generalizations are difficult to make in this area and much of the work will be specific in house work where a biotechnological process may be one step in an overall synthesis.

As for the future, it is difficult to predict the price of hydrocarbon feedstocks in relation to agriculturally derived feedstocks. The oil price is in a state of flux and may stay at a low level compared with its 1980s peak for some time. World sugar prices can vary wildly (*Figure 3*) and may not reflect the price paid by consumers in many areas (Brown, 1983) for example in the EEC. Recent world prices have been below the production cost of about US$250 a tonne for efficient producers. How long this will remain is uncertain and is difficult to predict (Anon, 1985) in an area where quotas, subsidies and trade restrictions abound. As for natural oils and fats, improvements in plant breeding could make these competitive with the other potential feedstocks. In such a situation feedstock flexibility could be an advantage for those processes where feedstock prices are an important part of manufacturing costs and where hydrocarbons would be acceptable if the price was right.

Acknowledgements

My thanks to many colleagues for useful discussions which contributed to this presentation, especially to Dr J.D.Linton for *Table 3*.

References

Akiyama,S., Suzuki,T., Sumino,Y., Nakao,Y. and Fukuda,H. (1973) Induction and citric acid productivity of fluoroacetate-sensitive, mutant strains of *Candida lipolytica. Agric. Biol. Chem.*, **37**, 879−884.

Anon (1985) Sugar sickens free traders. *The Economist*, March 2, 67.

Brink,L.E.S. and Tramper,J. (1985) Optimisation of organic solvent in multiphase biocatalysis. *Biotechnol. Bioeng.*, **27**, 1258−1269.

Brooks,J.D. and Meers,J.L. (1973) The effect of discontinuous methanol addition on the growth of a carbon-limited culture of *Pseudomonas. J. Gen. Microbiol.*, **77**, 513−519.

Brookes,I.K., Lilly,M.D. and Drozd,J.W. (1986) Stereospecific hydrolysis of *d,l*-menthyl acetate by *Bacillus subtilis*: Mass transfer-reaction interactions in a liquid-liquid system. *Enzyme Microb. Technol.*, **8**, 53−57.

Brown,O. (1983) Political and economic aspects of fermentation raw materials markets. *Chem. Ind.*, **3**, 95−97.

Champagnat,A., Vernet,C., Laine,B. and Filosa,J. (1963) Biosynthesis of protein-vitamin concentrates from petroleum. *Nature*, **197**, 13−14.

Cooney,C.L. (1979) Conversion yields in penicillin production: theory versus practice. *Process Biochem.*, **14**, 31−33.

Dale,B.E. and Linden,J.C. (1984) Fermentation substrates and economics. *Annu. Rep. Ferment. Proc.*, **7**, 107−134.

Dalton,H. (1980) Oxidation of hydrocarbons by methane monooxygenase from a variety of microbes. *Adv. Appl. Microbiol.*, **26**, 71−87.

Datar,R. (1986) Economics of primary separation steps in relation to fermentation and genetic engineering. *Process Biochem.*, **21**, 19−26.

Denyer,E. (1983) Political and economic aspects of the starch hydrolysates market. *Chem. Ind.*, **3**, 99−101.

Dimmling,W. (1976) Feedstocks for large-scale fermentation processes. In *Microbial Energy Conversion*. Schlegel,H.G. and Barnea,J. (eds), Erich Goltze K.G., Göttingen, FRG, pp. 499−514.

Dimmling,W. and Nesemann,G. (1985) Critical assessment of feedstocks for biotechnology. *Crit. Rev. Biotechnol.*, **2**, 233−285.

Drozd,J.W. (1980) Whole cell transformations. In *Hydrocarbons in Biotechnology*. Harrison,D.E.F., Higgins,I.J. and Watkinson,R. (eds) Heyden and Son on behalf of the Institute of Petroleum, London, pp. 75−83.

Drozd,J.W. and McCarthy,P.W. (1981) Mathematical model of microbial hydrocarbon oxidation. In *Microbial Growth on C1 Compounds. Proceedings of Third International Symposium, Sheffield, UK.* Dalton,H. (ed.), Heyden, London, pp. 360−369.

Dunnill,P. (1981) Biotechnology and industry. *Chem. Ind.* 7, 204−217.

Engelskirchen,K., Stein,W., Bahn,M., Schieferstein,L., Schindler,J. and Schmid,R. (1985) Verbessertes Verfahren zur Herstellung von Xanthomonas - Biopolymeren. European Patent 0058 364.

Fukui,S. and Tanaka,A. (1980) Production of useful compounds from alkane media in Japan. *Adv. Biochem. Eng.*, **17**, 1−35.

Gill,C.O. and Ratledge,C. (1972) Toxicity of *n*-alkanes, *n*-alk-1-enes; *n*-alkan-1-ols and *n*-alkyl-1-bromides towards yeasts. *J. Gen. Microbiol.*, **72**, 165−172.

Hydrocarbons as feedstocks for biotechnology

Habets-Crützen,A.Q.H., Brink,L.E.S., van Ginkel,C.G., de Bont,J.A.M. and Tramper,J. (1984) Production of epoxides from gaseous alkenes by resting-cell suspension and immobilised cells of alkene-utilising bacteria. *Appl. Microbiol. Biotechnol.*, **20**, 245−250.

Habets-Crützen,A.Q.H. and de Bont,J.A.M. (1985) Inactivation of alkene oxidation by epoxides in alkene- and alkane-grown bacteria. *Appl. Microbiol. Biotechnol.*, **22**, 428−433.

Habets-Crützen,A.Q.H., Carlier,S.J.N., de Bont,J.A.M., Wistuba,D., Schurig,V., Hartmans,S. and Tramper,J. (1985) Stereospecific formation of 1,2-epoxypropane, 1,2-epoxybutane and 1-chloro-2,3-epoxypropane by alkene-utilising bacteria. *Enzyme Microb. Technol.*, **7**, 17−21.

Hamer,G. (1979) Biomass from natural gas. In *Economic Microbiology*. Rose,A.H. (ed.) Academic Press, London, Vol. 4, pp. 315−360.

Hamer,G. (1981) Progress in fermentation technology resulting from single-cell protein process research and development. In *Advances in Food Producing Systems for Arid and Semi-arid Lands*. Academic Press, London, pp. 155−190.

Hamer,G. (1985) Impact of economic strategies on biotechnological developments. *Trends Biotechnol.*, **3**, 73−79.

Hamer,G. and Hamdan,I.Y. (1979) Protein production by micro-organisms. *Chem. Soc. Rev.*, **8**, 143−170.

Hamer,G., Hamdan,I.Y. Khamis,A.S. and Baroon,Z.H. (1980) Problems concerning the operation of high productivity biomass fermentations using volatile liquid hydrocarbon feedstocks. *Biotechnol. Bioeng.*, **22**, 995−1006.

Hou,C.T. (1984) Other applied aspects of methylotrophs. In *Methylotrophs: Microbiology, Biochemistry and Genetics*. Hou,C.T. (ed.), CRC Press, Inc., Boca Raton, Florida, pp. 145−166.

Humphrey,A.E. and Erickson,L.E. (1972) Kinetics of growth on aqueous-oil and aqueous-solid dispersed systems. *J. Appl. Chem. Biotechnol.*, **4**, 125−147.

Hyde,M. (1978) Chemical Insight No. 155, 1−3.

Kieslich,K. (1976) *Microbial Transformations*. John Wiley and Sons. George Thieme Publishers, Stuttgart.

King,P.P. (1982) Biotechnology. An Industrial View. *J. Chem. Technol. Biotechnol.*, **32**, 2−8.

Kroner,K.H., Hustedt,H. and Kula,M.R. (1984) Extractive enzyme recovery: economic considerations. *Process Biochem.*, **19**, 170−179.

Linton,J.D. and Buckee,J.C. (1977) Interactions in a methane-utilising mixed culture in a chemostat. *J. Gen. Microbiol.*, **101**, 219−225.

Linton,J.D. and Drozd,J.W. (1982) Interactions and communities in biotechnology. In *Microbial Interactions and Communities*. Bull,A.T. and Slater,J.H. (eds), Academic Press, London, Vol. 1, pp. 357−406.

Miall,L.M. (1980) Organic acid production from hydrocarbons. In *Hydrocarbons in Biotechnology*. Harrison,D.E.F., Higgins,I.J. and Watkinson,R. (eds), Heyden and Son Ltd., on behalf of The Institute of Petroleum, London, pp. 25−34.

Neidleman,S.L., Amon,W.F.,Jr. and Geigert,J. (1981) Preparation of epoxides and glycols from gaseous alkenes. United States Patent 4, 284, 723.

Pape,M. (1976) The competition between microbial and chemical processes for the manufacture of basic chemicals and intermediates. In *Microbial Energy Conversion*. Schlegel,H.G. and Barnea,J. (eds), Erich Goltze,K.G. Göttingen, FRG, pp. 515−530.

Ratledge,C. (1980) Microbial lipids derived from hydrocarbons. In *Hydrocarbons in Biotechnology*. Harrison,D.E.F., Higgins,I.J. and Watkinson,R. (eds), Heyden and Son Ltd., on behalf of The Institute of Petroleum, London, pp. 133−153.

Ratledge,C. (1983) Microbial oils and fats: an assessment of their commercial potential. *Prog. Ind. Microbiol.*, **16**, 119−206.

Ratledge,C. (1984) Biotechnology as applied to the oils and fats industry. *Fette Seifen Anstrichmittel.*, **86**, 379−389.

Rehm,H.J. and Reiff,I. (1981) Mechanisms and occurrence of microbial oxidation of long-chain alkanes. *Adv. Biochem. Eng.*, **19**, 175−215.

Seipenbush,R. and Blenke,H. (1980) The loop reactor for cultivating yeast on *n*-paraffin substrates. *Adv. Biochem. Eng.*, **15**, 1−40.

Shennan,J.L. and Levi,J.D. (1974) Growth of yeasts on hydrocarbons. *Prog. Ind. Microbiol.*, **14**, 1−57.

Stowell,J.D. and Bateson,J.B. (1983) Economic aspects of industrial fermentations. In *Bioactive Microbial Products 2. Development and Production*. Nisbet,L.J. and Winstanley,D.J. (eds), Academic Press, London, pp. 117−139.

Suzuki,T., Yamaguchi,K. and Tanaka,K. (1971) Effects of cupric ion on the production of glutamic acid and trehalose by *n*-paraffin-grown-bacterium. *Agric. Biol. Chem.*, **35**, 2135−2137.

Tramper,J., van der Plas,H.C. and Linko,P. (1985) *Biocatalysts in Organic Synthesis. Studies in Organic*

Chemistry 22. Elsevier, Amsterdam.

Van der Beek,C.P. and Roels,J.A. (1984) Penicillin production: biotechnology at its best. *Antonie van Leeuwenhoek,* **50**, 625−639.

Voelskow,H., Keller,R. and Schlingmann,M. (1985) Verfahren zur Verringerung der Viskosität von Fermentationsbrünen. German Patent DE 33 30 328 Al.

Wingard,L.B., Roach,R.P., Miyawaki,O., Egler,K.A., Klinzing,G.E., Silver,R.S. and Brackin,J.S. (1985) Epoxidation of propylene utilising *Nocardia corallina* immobilised by gel entrapment or adsorption. *Enzyme Microb. Technol.,* **7**, 503−508.

CHAPTER 8

The application of oils and fats in antibiotic processes

J.D. STOWELL

Pfizer Chemicals Division, Sandwich, Kent, UK

Introduction

Of the diverse range of carbon substrates potentially available for antibiotic fermentations, there is no doubt that carbohydrates feature most prominently in the published literature on the subject. One reason for this is that many of the publications, particularly those on physiological and biochemical aspects, are based on laboratory studies, and clearly carbohydrates provide a convenient basis for fundamental investigations. At production scale, where process economics become a major determining factor, the picture is somewhat different. At this scale of operation carbohydrates face serious competition from oils and fats as the major substrate of choice.

Oils were first used in antibiotic processes as carriers for anti-foams (Solomons, 1969). However, as the emphasis was placed on improving product concentrations, and fermentation cycles were extended to weeks rather than days, it soon became clear that oils and fats can provide a readily available, consistent and cost effective form of energy.

The initial objective of this chapter is to outline some of the points for and against oils and fats as substrates for antibiotic processes. An attempt will be made to summarize the range and nature of the materials available, to describe some of the special points and problems associated with their use, and to indicate possible future trends.

Obviously a subject as broad as this can be considered from many angles and, at best, only superficially. It is felt appropriate to represent the antibiotic producer's viewpoint and hence discussion will be confined to those points relevant to the production manager in deciding on his choice of carbon substrates.

The case for oils and fats

The choice of carbon substrate for a given antibiotic fermentation is a highly complex affair, requiring both technical and economic input. Ultimately the overall process economics play the major determining role. For example, for a single Pfizer antibiotic process, as many as 10 quite distinct carbon substrates have been, or are being, used for commercial production, dependent on geographical location of the production facility and prevailing economics. Many more have been evaluated at pilot scale.

Bader *et al.* (1984) described some of the key technical points favouring the choice of an oil as compared with a carbohydrate. Energetically, a typical oil contains approxi-

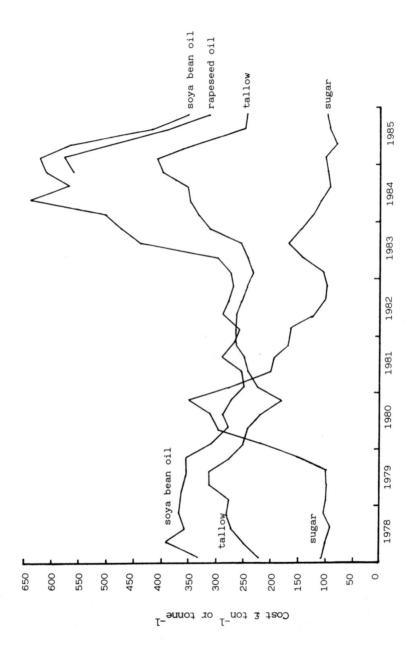

Figure 1. Relative commodity prices 1978 to date. All data are quarterly averages obtained from the Public Ledger. Soyabean oil and rapeseed oil — spot price London area, £/tonne, tallow — English Grade 2 *ex* works £/tonne, sugar — London daily raw price £/tonne.

mately 2.4 times the energy of glucose on a per weight basis. Glucose oxidation can be described by the following equation:

$$C_6H_{12}O_6 + 6O_2 \rightarrow 6CO_2 + 6H_2O + 670 \text{ kcal/mol glucose}$$
$$\text{kcal/kg glucose} = 3722$$
$$\text{kcal/mol } O_2 = 112$$

whereas for the oxidation of a typical oil, the energy value is:

$$C_{55}H_{108}O_9 + 77.5O_2 \rightarrow 55CO_2 + 54H_2O + 8100 \text{ kcal/mol oil}$$
$$\text{kcal/kg oil} = 8880$$
$$\text{kcal/mol } O_2 = 105$$

This fact should be taken into consideration when looking at relative commodity prices. It can be seen from *Figure 1* that, on an available energy basis, it would have been cheaper to have used sugar during the period 1978 to mid 1979 and late 1983 to 1985, whereas oil would have been the substrate of choice during the intervening period. Clearly, technical flexibility is a key to optimum process economics.

Oils are certainly preferable to carbohydrates on a volume basis. For example, it takes 1.24 litres of soyabean oil to add 10 kcal of energy to a fermenter, whereas it takes over 5 litres of glucose or sucrose to add the same amount of energy, assuming that the latter are added as 50% w/w solutions. Maximizing tank fill is an important aspect of any commercial fermentation process and oil permits better fine tuning of this. A greater void volume must be available when a carbohydrate is used in order to allow response to a sudden reduction in the residual nutrient level.

Oils have built in anti-foam properties which can also give them a competitive edge over carbohydrates. This is particularly so when ultrafiltration is involved in product recovery, as many proprietary anti-foams can give rise to problems of membrane fouling. However, under other circumstances it is not desirable to use oil as a sole source of anti-foam as this can easily lead to gross over usage.

In defence of carbohydrates, it should be noted that the problems of fermenting a two-phase system are not inconsequential. In addition, as noted above, oil is less oxidized than carbohydrate and hence, for a given energy input to a fermentation, the oil has a higher oxygen demand (+7%). This is a very real consideration for many antibiotic processes where oxygen can be the limiting nutrient.

Commercial aspects of oils and fats

Worldwide availability

Table 1 lists the world production of oils and fats for the 1981/82 season. It can be seen that soyabean oil accounted for some 22.6% of the total, more than twice the tonnage of the next most important commodity. The trend during the decade 1970−1980 was one of gradual growth in production, with the developing countries playing an increasingly important role (*Table 2*). In the period 1970−71 to 1980−81 overall production increased from 42 432 to 57 245 million tonnes (Mt), an increase of 35%. During this period soya increased by 98%, palm by 84%, rapeseed by 54% and sun-

Table 1. World production of oils and fats in 1981/82[a].

Material of origin	Oil/fat production (in 1000 tonnes)	Percentage of total
Soyabeans	13 154	22.6
Tallow	6038	10.4
Palm oil	5414	9.3
Sunflower seed	5075	8.7
Butter (fat basis)	4962	8.5
Rapeseed	4115	7.1
Lard	3804	6.5
Cottonseed	3506	6.0
Copra	3238	5.6
Peanuts	3154	5.4
Olive oil	1249	2.1
Fish	1234	2.1
Palmkernel	735	1.3
Sesameseed	651	1.1
Flaxseed	622	1.1
Corn oil	525	0.9
Castorbeans	393	0.7
Safflowerseed	242	0.4

[a]From Leysen, 1982.

Table 2. Trend in world oils and fats production, 1970−1980[a].

	Average production 1970−1971	Average production 1975−1976	Average production 1979−1980	Growth rate 1970−1980 (% per annum)
World total	43 080	50 100	59 990	3.7
Developed	25 740	28 250	34 990	3.3
Developing	17 340	21 840	25 010	4.3

[a]From Hancock, 1982.

flower by 30%. Production of most of the other oils remained almost static (Pigden, 1983).

UK situation

The UK has a well developed seed crushing and oil processing industry based mainly in Merseyside/Manchester, East London and Hull, and represented by SCOPA (the Seed Crushers and Oil Processors Association).

The increasing importance of rapeseed as a crop can hardly have escaped the attention of most of us. Almost unheard of in the UK some 20 years ago, in 1985 for the first time the area under cultivation with rapeseed exceeded that with potatoes. This trend is indicated in *Figure 2*. A 6-fold increase in rapeseed oil production was recorded in the UK in the 10 years to 1981. *Table 3* shows a more than doubling in rapeseed processing during the period 1980−1984 whilst at the same time the UK soyabean crush halved. This owes much to selective breeding and to the EEC Common Agricultural Policy (CAP) as noted below.

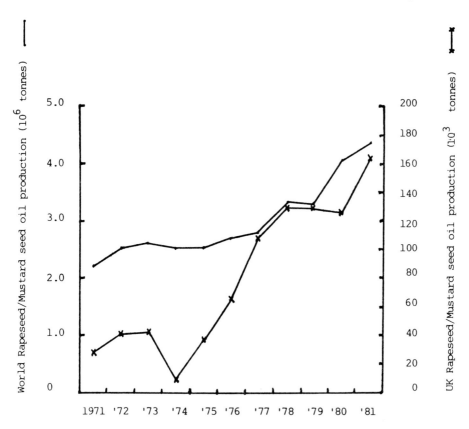

Figure 2. UK and world rapeseed/mustard seed oil production, 1971−1981 (re-plotted from Pigden, 1983).

Table 3. Share of oilseeds crushed in the UK.

Material	Percentage of oilseed in UK crush	
	1980	*1984*
Soyabeans	62%	33%
Rapeseed	20%	52%
Palm kernels and tropical seeds	7%	4%
Sunflower seed	6%	Not quoted
Maize germ	3%	2%
Linseed and castorbeans	2%	5%
Other	Not quoted	5%
Total crush	1.8×10^6 tonnes	1.2×10^6 tonnes
Crude oil produced	0.48×10^6 tonnes	0.38×10^6 tonnes

Data taken from the SCOPA booklet (Anon. 1980, with additional data for 1984).

Production of oils and fats

The processing of most oilseeds follows the route summarized in *Figure 3*. The point of entry in the process is dependent on the oil content of the seed. High oil seeds, for example rapeseed (40−45% oil), are expelled prior to solvent extraction, whereas the

143

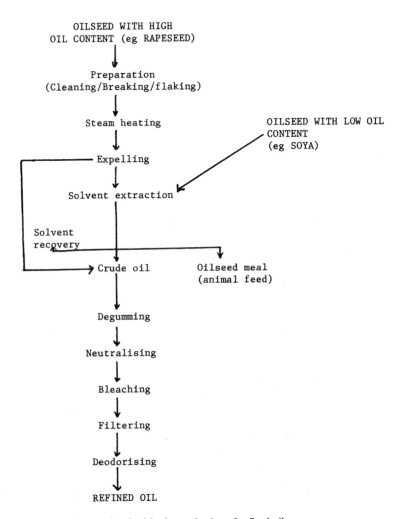

Figure 3. Summary of stages involved in the production of refined oils.

lower oil content seeds, for example soya (19−20% oil), are solvent extracted as the first step. Processing adds a substantial value to the product and hence, for the fermentation technologist seeking a cheap carbon substrate, it is important to identify the least pure oil acceptable from a technical point of view.

Influence of the EEC/world economy on price and availability

It has already been shown (*Figure 1*) that during the 1983−84 season soyabean oil prices more than doubled. Equally rapidly the prices halved during 1985. There was a similar trend in sugar prices between 1979 and 1981. These examples serve to illustrate the volatility of commodity prices, where an imbalance of supply and demand, either actual or predicted, can have a rapid and drastic effect. Food use accounts for over 80% of the consumption of oils and fats and hence the industrialist wishing to use these commodities has little or no hold over prices.

The EEC is far from self-sufficient in oilseed production. In 1980−81 the degree of self-sufficiency was 18%. This rose to 33% in 1985, but will fall to 31% with the entry of Spain and Portugal in 1986. (Data from The Public Ledger; Anon, 1986.) Rexen and Munck (1984) quoted a figure of 22% self-sufficiency for 1982, but they pointed out that 79% of vegetable oil consumed within the EEC is extracted domestically. In 1982 the EEC produced 3.8 Mt of oil seed, imported 140 Mt and exported only 25 thousand tonnes.

As might be expected the EEC is self-sufficient in olive oil and it also produces all its rapeseed oil requirements. The prediction is that rapeseed oil production will increase to 4 Mt by 1990 (from 2.7 in 1982) and sunflower oil, the other major indigenous crop, to 1.4 Mt (from 0.74 in 1982).

The EEC CAP is designed to improve the Community's self-sufficiency with regard to vegetable oils and to this end the seed crushers have been receiving a subsidy. For rapeseed this has been as high as £170/tonne, but in recent years a world shortage in soya has led to a narrowing in the gap between EEC and world prices and hence in a reduction in the requirement for subsidy. As of May 1985 the subsidy was £42 – £57/tonne and, due to EEC financial constraints, its future uncertain. Crushers have pointed out that this is insufficient for them to maintain a profitable business. Indeed, Unilever has recently reduced substantially its crushing capacity within Europe.

World overproduction is currently having a destabilizing effect on the price of oils and fats. The example of Malaysia and Indonesia serves to illustrate the nature of the problem. In these countries reduced revenue from petroleum and rubber has led to a massive switch to palm oil production in order to attract foreign exchange. Malaysian palm oil exports increased from 2.5 Mt in 1981 to 4.0 Mt in 1985 and a figure of 6.5 Mt is predicted by 1990. As much as 95% of this is exported after refining (see 'Financial Times', July 24, 1985).

The consequence of this increased production has been that palm oil prices are now at an all time low and world stocks at almost 1 Mt. This of course is putting the pressure on US soya, which in turn affects the profitability of rape within the EEC.

The overall message from the foregoing is that the antibiotic producer has little influence over general trends in oil and fat commodity prices and clearly, as already noted above, efforts should be directed towards achieving technical flexibility.

Chemical composition of oils and fats

The terms 'oil' and 'fat' are essentially interchangeable, both consisting mainly of glyceryl esters of fatty acids, or triglycerides. Commonly those triglycerides which are liquid at ambient temperature are called oils and those solid or semi-solid, are called fats.

A triglyceride is a condensation product of one molecule of glycerol with three molecules of fatty acids:

$$
\begin{array}{llll}
CH_2OH & + R_1COOH & CH_2OOCR_1 & \\
| & & | & \\
CHOH & + R_2COOH & \rightarrow CH\text{-}OOCR_2 & + 3H_2O \\
| & & | & \\
CH_2OH & + R_3COOH & CH_2\text{-}OOCR_3 & \\
\text{Glycerol} & \text{fatty acids} & \text{triglyceride} & \text{water}
\end{array}
$$

Table 4. Fatty acid composition of a range of oils and fats.

		Soyabean oil	Rapeseed oil[a]	Rice bran oil	Cottonseed oil	White grease	Crude tallow	Norwegian fish oil
Saturated acids								
Myristic	14:0	–	–	0.4	0.5	0.9	1.9	3.7
Palmitic	16:0	10.4	4.8	8.4	16.5	15.6	20.5	8.8
Stearic	18:0	3.4	0.9	2.3	2.0	5.3	15.4	1.5
Arachidic	20.0	–	–	2.4	–	–	–	–
Behenic	22:0	–	–	2.1	–	–	–	–
Lignoceric	24:0	–	–	–	–	–	–	3.7
Unsaturated acids								
Myristoleic	14:1	–	–	–	–	–	–	–
Palmitoleic	16:1	–	–	0.3	0.3	10.2	4.4	5.1
Oleic	18:1	24.6	55.9	28.0	23.1	55.1	44.9	7.9
Linoleic	18:2	53.9	25.6	45.5	49.1	10.2	4.7	1.8
Linolenic	18:3	7.7	12.8	3.7	1.9	2.1	1.8	7.9
Arachidonic	20:4	–	–	–	–	–	–	22.1
Cetoleic	22:1	–	–	–	0.6	–	–	–
Erucic	22:1	–	–	–	0.9	–	–	10.6
Docosahexanoic	22:4	–	–	–	–	–	–	1.8
Nervonic	22:1	–	–	–	–	–	–	10.0
Unknown	C22-C26	–	–	–	–	–	–	14.8

Data were obtained by gas – liquid chromatography by S.V. Hammond (personal communication). The values agree closely with those in the published literature.
[a]Zero erucic.

The physical and chemical properties of individual oils and fats are largely determined by their fatty acid composition. The fatty acids can either be saturated or unsaturated, that is containing one or more double bond:

$$CH_3 (CH_2)_n COOH \qquad CH_3 (CH_2)_a CH=CH(CH_2)_b COOH$$
Saturated fatty acid   Mono-unsaturated fatty acid

Most naturally occurring triglycerides are mixed in that they have two or three different fatty acid chains per molecule, and these may be distributed randomly or semi-randomly around the glyceryl moiety.

Materials with a high degree of unsaturation tend to be more reactive than the predominantly saturated fats, due to the presence of the double bonds.

Table 4 gives the typical fatty acid distribution for a range of commercially important oils and fats. It can be seen that all the fatty acids listed have even numbers of carbon atoms and C_{18} mono- and dienoic acids (i.e. 1 and 2 double bonds) are the most predominant in the oils of vegetable origin.

The rapeseed oil referred to in *Table 4* is the variety lacking erucic acid which is by far the most common type now grown. Up to about 1950 all rapeseed oil contained about 45% erucic acid (C22:1). This was shown to be poorly metabolized at least by the rat, and was also implicated in fat accumulation particularly around the heart muscle. In addition, sulphur-containing components of naturally occurring rapeseed, the glucosinolates, possessed anti-thyroid activity. A major triumph of post-war plant breeding has been to produce an erucic acid- and glucosinolate-free rapeseed, suitable both for human consumption and for animal feed use.

All oils and fats contain small amounts of non-glyceride components. In the case of crude vegetable oils this is less than 5% whilst for refined oils it is below 2%. Phosphatides are the major non-triglyceride constituent of crude oils. These are essentially di-glycerides with an esterified phosphoric acid attached to the third carbon of the glycerol skeleton. Soya has the highest phosphatide content, at about 1.8% average.

Most oils contain some carbohydrate, usually associated with the phosphatides. Raffinose and pentosans are particularly prevalent in crude cottonseed oil. Protein degradation products are present in crude oils to a greater or lesser extent depending on the oil extraction method. Other trace components include sterols, waxes, hydrocarbons, fat-soluble vitamins A, D and E, anti-oxidants and minerals. In addition to these, oils contain minute quantities of pigments, particularly carotenoids, and flavour compounds which bestow unique characteristics on each oil.

From an antibiotic producer's point of view the non-triglyceride components of an oil can make an important contribution to its suitability as a fermentation substrate. For example, phosphatides are an important source of phosphorus, essential for sustaining growth during a prolonged fermentation cycle. The value of anti-oxidants and waxes is noted in the section below.

Handling of oils and fats at production scale

Key technical areas requiring the production manager's attention include Quality Control testing, storage, sterilization, metering of oil into fermenters and the derivation of an optimum oil feed strategy.

Quality control

All raw materials to be used in the manufacture of an antibiotic must be rigorously tested against a purchase specification, and each lot of material bought in must be unambiguously identifiable for possible future reference. Hence, after delivery, individual batches must be quarantined pending the results of Quality Control analyses. A typical purchase specification for oils and fats would include definition of the following parameters:

(i)   physical appearance;
(ii)   saponification value;
(iii)   iodine value;
(iv)   acid value;
(v)   moisture content;
(vi)   fatty acid content (by gas–liquid chromatography).

A simple physical inspection of the material when liquid would indicate presence of undesirable extraneous matter. In the case of fats this would require initial heating. The saponification value (SV) is a measure of the average fatty acid chain length. It is defined as the milligrams of potassium hydroxide required to hydrolyse 1 g of material, and the higher the SV the greater the proportion of low chain length fatty acids and vice versa.

The iodine value is a measure of the average degree of unsaturation of the fatty acids, and the acid value defines the free fatty acid content. These tests are all traditional, and well defined in the literature (see Grimstone, 1958; Swern, 1964). In conjunction with gas−liquid chromatography to quantify the fatty acid contact and type (see Kuksis, 1978), they permit a comprehensive qualitative analysis of the oil prior to use.

Storage

Storage of vegetable oils for a year or more is normal. Indeed, most vegetable oils are derived from crops which are only harvested annually. Given this, the consistency of material used during the season is remarkably good. A key reason for this is the presence in vegetable oils of natural anti-oxidants. The most important and widely distributed of these are the tocopherols, that is fat soluble vitamin E.

Tocopherols are present in crude soyabean oil at 0.17% and in refined soyabean oil at 0.1%. Animal fats contain much lower levels of natural anti-oxidants. For example, beef tallow contains only 0.001% tocopherols. In addition, animal fats must be heated to maintain them in liquid form, 40°C being typical. Consequently anti-oxidants must be added to animal fats in order to prevent spoilage. These can either be natural products such as tocopherol concentrates and selected vegetable oils, or chemically synthesized products, of which butylated hydroxyanisole (BHA), butylated hydroxytoluene (BHT) and propyl gallate are probably most important. Combinations of these are more effective than individual materials and ascorbic, phosphoric and citric acids are used as synergists, i.e. to improve the anti-oxidant capabilities.

Many oils, particularly crude vegetable oils, contain a certain amount of solid matter or sludge. Crude rapeseed and rice bran oils are notable in this respect. If this sludge is of no benefit to the fermentation in question, then it is desirable to remove as much as possible by settling and decanting off the oil. Solid matter can seriously impair the performance of sterilizers and metering equipment — see below.

Sterilization

On the face of it sterilization of a liquid feed into a production fermenter should present few problems for a competent chemical engineer. However, in reality this is certainly not the case. Bader *et al.* (1984) recently concluded in an interesting paper on the subject that lack of adequate oil sterilization may well represent a major contamination source in the fermentation industry. Practically, sterilization of oil at production scale can be accomplished either by heat or by filtration. Recent developments in membrane filter technology have made the latter a viable proposition, but by far the majority of traditional fermentation plants are undoubtedly still using heat sterilization.

The bacterial challenge to the sterilization system will clearly be variable, and is often difficult to determine. Gram-negative organisms are destroyed even by relatively mild heat. Most are completely killed after 5 min incubation in dry oil at 120°C. Hence, realistically only Gram-positive contamination can be attributed to inadequacies in oil sterilization systems. Of course the integrity of the line downstream of the sterilizer may be in doubt, in which case feeding oil hot into the fermenters would be desirable.

Oil sterilizers can either be batch or continuous, with or without live steam injection. Heat requirements for sterilization of dry oil tend to be similar to those for dry

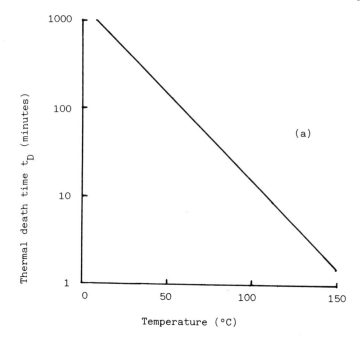

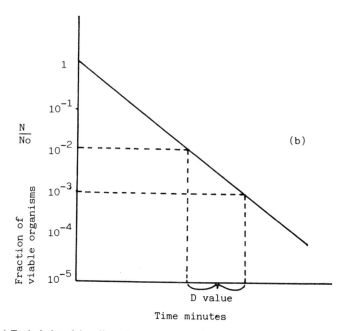

Figure 4. (a) Typical plot of the effect of temperature on thermal death time, i.e. the time taken to completely kill a given microorganism under a given set of conditions. **(b)** Plot showing the calculation of D value, i.e. the time taken to achieve a 90% kill, at a given temperature.

conditions rather than wet conditions, that is very much higher temperatures and longer holding times are required.

Clearly live steam injection, or sterilization of oil – water emulsions would reduce the heat requirements. However, the presence of water in oil can have an adverse effect on metering equipment and, for certain fermentations, the feeding of an emulsion can create severe foaming problems.

Batch sterilization is preferred to continuous sterilization in that it guarantees a minimum residence time in the sterilizer. However, it is more difficult to automate batch sterilizers and, if live steam is not involved, it can be difficult to sterilize the head space above the oil and associated ancillary equipment.

The requirement of the sterilization system is a complete kill of all contaminating microorganisms. Indeed, an inoculum of 1 – 10 contaminating organisms can, under favourable circumstances, completely take over an antibiotic fermentation.

In general, thermal death of microorganisms conforms to the equation:

$$\mathrm{Ln}\left(\frac{N}{N_0}\right) = -kt$$

where Ln N/N_0 is the fraction of viable organisms surviving after heat treatment for time t, and k is the thermal death rate constant (Richards, 1968). The thermal death time t_D is the time taken for the viable count to reach zero, whilst the D value is the time taken, at a given temperature, for the count to reach $1/10$ of its previous value. Typical plots are given in *Figure 4*.

Bader *et al.* (1984) obtained a straight line plot for Log D value against temperature for *Bacillus subtilis* spores in lard oil of number 2 grade and soyabean oil. A continuous point of use sterilizer, of a shell and tube type, that is with a steam jacket surrounding a spiral tube, was designed on the basis of the measured D value for *B. subtilis* in lard oil. The design took into account the heat transfer coefficient, and assumed a minimum residence time of half the mean residence time. However, the sterilizer was subsequently found to be ineffective in sterilizing soyabean oil. Further investigation revealed the presence of *Bacillus macerans* spores which had a D value 100 times greater than that of the *B. subtilis* spores. The D value was found to be very much dependent on whether the spores were initially wet, in which case their D value was similar to that for *B. subtilis* or dry, in which case the higher value was obtained. The more detailed knowledge subsequently enabled the successful design of the sterilizer.

An example taken from the Fermentation Group at Pfizer Ltd., Sandwich, serves to illustrate the complexity of the problem. A continuous sterilizer of the type shown in *Figure 5* is used for sterilizing oil for production fermenters. Initially the system was set up without the plate and frame pre-heater. The oil system had been suspected as a source of Gram-positive contamination, and laboratory tests had shown that spores of the contaminant in question, *Bacillus circulans*, could survive for up to 40 min at 150°C (*Table 5*).

The mean and minimum residence times of oil in the sterilizers were determined by introducing a slug of BHT on the non-sterile side and monitoring its reappearance after one sterilizer and on the 'sterile' side using gas – liquid chromatography.

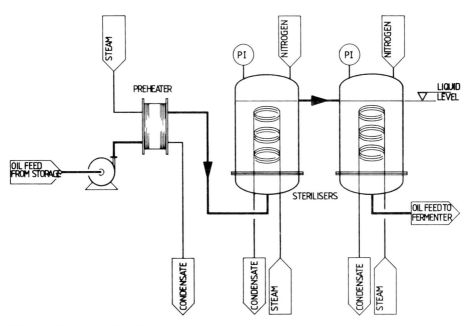

Figure 5. Continuous oil sterilization system.

Table 5. Survival of *Bacillus circulans* spores in rapeseed oil.

Temperature	*Time (min)* *Control*	*10*	*20*	*30*	*40*	*50*	*60*	*90*
120°C	+	−	−	+	+	−	○	○
130°C	+	−	−	+	+	−	○	○
140°C	+	+	+	+	+	+	○	○ .
150°C	+	+	+	+	+	○	○	○

The culture was grown on sporulating agar (Arret and Kirshbaum, 1959) and harvested into rapeseed oil. The spore suspension (1 cm^3) was then added to tubes each containing 10 cm^3 of oil. These were incubated for the appropriate time in an electrically heated block. Subsequent to this they were transferred to nutrient broth (100 cm^3), shaken for 2 days at 28°C, subcultured onto Tryptone agar plates and inspected for evidence of growth after a further 2 days' incubation.
+ = evidence of growth.
○ = no growth.
− = not determined.

The experiment was carried out both before and after cleaning the system out with a caustic wash. The results (*Figure 6*) indicated that when the system was clean an adequate minimum residence time was achieved. However, the design of the system was clearly unsatisfactory in that the minimum residence time was only some 25% of the mean, and when the system became fouled the minimum sterilization criteria were not met. Under these circumstances the sludge acted as an insulator on the heat exchanger such that the desired sterilization temperature, 150°C, was not achieved. Removal of the sludge prior to sterilization would require substantial capital investment and in any case the majority of the solid matter was only actually formed on heating the oil.

In order to try to improve the integrity of the system a pre-heater was installed which

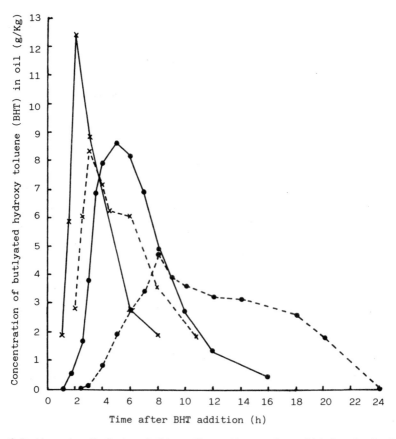

Figure 6. Residence time distribution of oil in sterilizers without pre-heater (a) before cleaning (x), and (b) after cleaning (●). The solid line (—) shows appearance of butylated hydroxy toluene (BHT) after the first sterilizer, and the hatched line (----) appearance at the exit of the second sterilizer. **Before cleaning**. Mean residence time: 1st pot 3.4 h; 2nd pot 2.15 h; entire system 5.5 h. Minimum residence time for entire system was less than 2 h. **After cleaning**. Mean residence time: 1st pot 6.3 h; 2nd pot 5.65 h; entire system, 11.95 h. Minimum residence time for entire system was between 2.5 and 3.0 h.

heated the oil up to 150°C prior to the first pot. Unfortunately this had the effect of reducing the minimum residence time (*Figure 7*) although it is debatable as to whether the actual sterilization conditions were worse, since without the pre-heater much of the oil in the first pot was not at a sufficiently high temperature. The reduction in minimum residence time was probably due to the lower viscosity hot oil creating mixing as it entered the bottom of the first pot.

Rice bran oil was used in a different system and in this case the sediment was removed by centrifugation prior to sterilization in order to prevent fouling of the pots. This resulted in excessive foaming of the fermentation broths. It was subsequently found that waxes in rice bran oil had a beneficial effect in controlling foam, and the situation was rectified by heating the oil to dissolve the waxes prior to centrifugation.

The take home message from the above is that the integrity of each sterilization system should be thoroughly tested before it can be ruled out as a source of contamination.

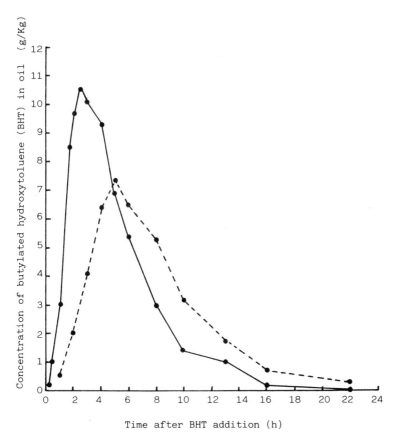

Figure 7. Residence time distribution of oil in sterilizers after installation of pre-heater. The solid line (—) shows appearance of BHT after the first sterilizer and the hatched line (-----) appearance at the exit of the second sterilizer. Mean residence time: 1st pot 4.54 h; 2nd pot 2.28 h; entire system, 6.82 h. Minimum residence time for entire system was less than 0.75 h.

Each oil has its own characteristics which make it impossible to predict performance of the system from one material to another.

Metering of oil into fermenters

Accurate metering of oil into fermenters is vital if the fermentation process in question is to be optimized. The systems used fall broadly into three categories, namely:

(i)   feeding of separate individual measured quantities (shots),
(ii)   metering pumps, and
(iii) a pressurized main plus in line flow monitoring equipment.

Discrete shots, added either manually or automatically, are fine if the primary use of the oil is as an anti-foam. However, continuous addition is preferred if the oil is to be the major carbon source. Metering pumps provide an accurate means of addition but there is often a question mark over their integrity from a sterility point of view. Bader *et al.* (1984) described a continuous point of use sterilizer of shell and spiral tube type which would overcome this point, but this would only be successful if the feed rate is to be constant at a pre-determined value.

Hence probably the most effective means of oil addition is to use a pressurized main in conjunction with flow monitoring equipment. Magnetic flow meters which rely on conductivity are unsuitable owing to the non-conducting nature of oils. However, the development of flow meters has in itself become something of a growth industry and many suitable devices exist based on principles such as turbines, ultrasonics, mass flow and vortex shedding.

Development of an optimum feed strategy

The main objectives when optimizing a feed strategy are to maximize productivity and to minimize costs. These may require different approaches (Stowell and Bateson, 1983). Traditionally feeds have been optimized on the basis of residual nutrient levels. This remains a viable option for many antibiotic processes where the time taken to perform the analysis is short, relative to the overall fermentation time. More recently, the development of accurate and reliable equipment for measuring fermenter 'off gas' has permitted optimization of carbon feeds on the basis of RO_2/RCO_2, the rate of oxygen uptake/CO_2 evolution or RQ, the respiratory quotient. For example, Cooney and Moo (1982) reported optimization of the glucose feed to a penicillin fermentation on this basis. This principle can equally well be applied to processes using oils and fats as carbon substrates.

Some physical and physiological considerations relating to oils and fats in fermenters

Residual oil in the fermenter

The limiting residual level for an oil in a fermenter can be substantially higher than that for a carbohydrate, which often approximates to zero. This can either be due to the physical limitation of oil mass transfer in the fermenter, or due to the inability of

Table 6. Comparison of the fatty acid spectrum of input material and final residual oil for a typical antibiotic fermentation.

	(a) Input material Soyabean oil		(b) Input material Animal fat	
	Input	Final residual oil	Input	Final residual oil
Saturated acids				
Myristic	0.5	–	1.5	1.0
Palmitic	6.7	5.3	24.6	27.8
Stearic	–	–	27.9	40.1
Unsaturated acids				
Palmitoleic	–	–	3.2	1.3
Oleic	28.2	29.3	32.3	23.2
Linoleic	47.8	49.5	7.2	4.5
Linolenic	3.3	2.6	1.4	1.4
Unknown	7.9	7.4	2.0	0.7

The residual oil was recovered by ether extraction and all assays were performed by g.l.c.

the organism to metabolize certain components, of the input material. For example, *Table 6* gives a comparison between the fatty acid spectrum of input material and final residual oil for a Pfizer antibiotic fermentation.

It can be seen that when soyabean oil was the substrate, the composition of the residual oil after fermentation was very similar to that of the oil added. Indeed, when the oil was extracted from the final broth and fed to a subsequent laboratory fermentation, the performance of that fermenter was equivalent to control. This strongly suggests a physical limitation to oil mass transfer in the fermenter.

However, when animal fat was fed there was a preferential utilization of the unsaturated fatty acids, with a net accumulation of stearic acid. This could either be due to better dispersion of the predominantly unsaturated fats in the fermenter or due to some metabolic preference for the unsaturated fatty acids.

A high level of residual oil is undesirable in that it tends to increase viscosity and hence reduce oxygen transfer efficiency ($K_L a$) in the fermenter. The situation can be improved by the addition of surfactants, that is, materials which contain both polar and non-polar groups and hence improve the stability of oil−aqueous emulsions.

The hydrophilic−lipophilic balance (HLB) is a numerical value which can be determined experimentally for a given oil−aqueous mixture. This can then be matched to a surfactant having the same value, such that addition of the surfactant will optimize the stability of the emulsion.

In vitro laboratory studies can be undertaken to determine the optimum surfactant for a given process and fermentation studies can ensure no adverse effect of the material on performance. However, it is often difficult to prove the value of the surfactant at production scale where mass transfer characteristics will be different from those in the laboratory, and where minor reductions in residual oil may be difficult to detect.

Nevertheless, the scope for cost savings is considerable. If it is assumed that a typical fermenter delivers 100 tonnes of broth with a residual oil of 2%, then for a plant running 100 fermentations *per annum* and using oil at £400 per tonne the value of the oil wasted would be £80 000, clearly worth minimizing.

Choice of material

The relative costs of different oils and fats have already been noted. However, overriding technical considerations may determine the final choice of material for a given process.

Figure 8 shows the performance of a Pfizer antibiotic process with a range of oils and fats. A good correlation is evident between performance, that is antibiotic production rate, and the percentage of the major unsaturated fatty acids in the material. The physical/physiological reasons for this are not entirely clear. It is evident that on a purely technical basis glycerol trioleate would be the material of choice. However, for this particular process, taking into account both technical and economic aspects, soyabean oil or rapeseed oil are the preferred substrates, at least in the UK. Glycerol trioleate is, however, known to be used for some antibiotic fermentations where substrate purity is the key consideration.

The difference in performance given by the different oils highlights the importance of performing laboratory studies with the same material as is used at production scale.

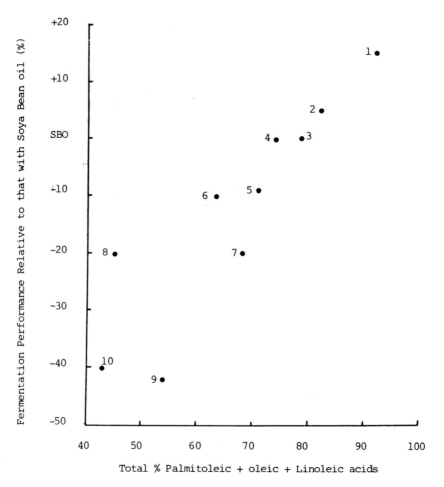

Figure 8. Laboratory fermentation performance of a Pfizer antibiotic process, using a range of oils and fats. The performance compared with that with soyabean oil is plotted against the percentage of the major unsaturated fatty acids in the material. Substrates 5–10 were solid at ambient temperature and a trace heated feed line was used. **1.** Glycerol trioleate; **2.** Rapeseed oil; **3.** Soyabean oil; **4.** Rice bran oil; **5.** Mixed vegetable oil; **6.** Animal fat (from France). **7.** Blended animal fat (Croda Blend 1); **8.** Animal fat (from Pure Lard); **9.** Crude tallow (Croda); **10**, Animal fat (Croda).

Metabolism of oils and fats

It is clearly beyond the scope of this chapter to discuss in any detail the metabolism of oils and fats. Suffice it to say that the main product of the β-oxidation of fatty acids (*Figure 9*) is acetyl-CoA and this is itself a precursor in the formation of many antibiotics. In fact, there are many similarities between fatty acid biosynthesis and the biosynthesis of the tetracycline antibiotics which are based on an 18 carbon ring structure formed by the condensation of malonyl-CoA units.

Future trends

It can be deduced from the foregoing that oils and fats have a well established role as carbon substrates for antibiotic fermentations. Future applications, both qualitative

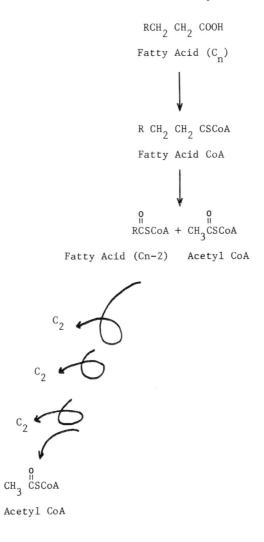

RCH$_2$ CH$_2$ COOH

Fatty Acid (C$_n$)

R CH$_2$ CH$_2$ CSCoA

Fatty Acid CoA

$$\overset{O}{\overset{\parallel}{RCSCoA}} + \overset{O}{\overset{\parallel}{CH_3CSCoA}}$$

Fatty Acid (Cn-2)   Acetyl CoA

C$_2$

C$_2$

C$_2$

$$\overset{O}{\overset{\parallel}{CH_3 \ CSCoA}}$$

Acetyl CoA

TCA Cycle

Figure 9. β-Oxidation of fatty acids to generate acetyl-CoA.

and quantitative, will be determined largely by economic considerations. Predicting market trends is obviously highly speculative but extremely important in dictating research and development priorities.

The EEC will certainly seek to become more self-sufficient in oils and fats, and to this end Rexen and Munck (1984) predict substantial increases in both rapeseed and sunflower production by the end of the century. The success of this programme will be very much dependent on the support of the EEC Common Agricultural Policy. The likely increase in palm oil production has also been referred to, and superimposed on all this is the certainty that the USA will not wish to see substantial reductions in its soya crop.

We have witnessed the ability of the plant breeder to respond to market forces in radically manipulating the rapeseed crop. This genetic manipulation is ongoing, with efforts now being concentrated on increasing the linoleic acid content of the oil for nutritional purposes, and reducing the linolenic acid content for technical use (Robbelen, 1984). Similarly a linolenic acid free soyabean oil is being sought (Smith, 1984).

Further into the future, the microbial production of oils and fats is receiving significant attention. Yeasts and moulds with up to 70% lipid content have been reported, and a 25% conversion efficiency is seen as a realistic target (Ratledge, 1984).

It can almost be taken for granted that bioreactors will be developed with improved mixing and oxygen transfer capabilities, thus overcoming the oxygen limitation which can be a particular feature of oil-fed antibiotic processes. Improved methods for liquid sterilization are also being developed, based on hydrophobic membrane technology and these lend themselves to oil applications.

Manufacturers currently producing antibiotics from carbohydrate feed stocks may well find that oils and fats represent a cost-effective alternative. Similarly, those seeking to develop new antibiotic processes should include oils as substrates in the initial screen. When successive generations of a particular culture are selected on the basis of their efficiency with one particular substrate, it can become increasingly difficult to change to a new substrate.

Acknowledgements

I am very much indebted to my colleagues within the Fermentation Group at Pfizer Ltd., Sandwich, in particular Nigel Cartwright, Brian Gedge, Steve Hammond, Nick Shimmin and Brian Woodland, who provided many of the results included in this article. I would also like to thank Peter Goodwin of Unichema for providing information on the supply and production of oils and fats in the UK market.

References

Anon. (1980) *SCOPA Seed Crushing and Oil Processing in the UK.* Libra Press, London.

Anon. (1986) *The Public Ledger and Daily Freight Register.* Turret Wheatland, Watford, February 8, p. 4.

Arret,B. and Kirshbaum,A. (1959) A rapid disc assay method for detecting penicillin in milk. *Milk Food Technol.*, **22**, 329–331.

Bader,F.G., Boekeloo,M.K., Graham,H.E. and Cagle,J.W. (1984) Sterilization of oils: data to support the use of a continuous point-of-use sterilizer. *Biotechnol. Bioeng.*, **26**, 848–856.

Cooney,G.L. and Moo,D.G. (1982) Application of computer monitoring and control to the penicillin fermentation. In *Computer Applications in Fermentation Technology.* Society of Chemical Industry, London. pp. 217–225.

Gunstone,F.D. (1956) *An Introduction to the Chemistry of Fats and Fatty Acids.* Chapman and Hall, London.

Hancock,R.F. (1982) The changing role of developing countries in the world economy of fats and oils; effects on the EEC. *Chem. Ind.*, **13**, 439–445.

Kuksis,A. (1978) Separation and determination of structure of fatty acids. In *Handbook of Lipid Research. Fatty Acids and Glycerides.* Plenum Press, London.

Leysen,R. (1982) Soya oil consumption trends in the EEC. *Chem. Ind.*, **13**, 428–431.

Pigden,W.J. (1983) World production and trade of rapeseed and rapeseed products. In *High and Low Erucic Acid Rapeseed Oils.* Kramer,J.K.G., Sauer,F.D. and Pigden,W.J. (eds), Academic Press, London pp. 21–59.

Ratledge,C. (1984) Microbial oils and fats—an overview. In *Biotechnology for the Oils and Fats Industry.*

Ratledge,C., Dawson,P. and Rattray,J. (eds), The American Oil Chemists' Society, Champaign, USA pp. 119−127.

Rexen,F. and Munck,L. (1984) *Cereal crops for industrial use in Europe*. EEC Commission Report (EUR 9617 EN).

Richards,J.W. (1968) Introduction to industrial sterilization. In *Thermal Sterilization of Liquids*. Academic Press, London, pp. 67−75.

Robbelen,G. (1984) Changes and limitations of breeding for improved polyenoic fatty acids content in rapeseed. In *Biotechnology for the Oils and Fats Industry*. Ratledge,C., Dawson,P. and Rattray,J. (eds), The American Oil Chemists' Society, Champaign, USA, pp. 97−105.

Smith,K.J. (1984) Genetic improvement of soyabean oil. In *Biotechnology for the Oils and Fats Industry*. Ratledge,C., Dawson,P. and Rattray,J. (eds), The American Oil Chemists' Society, Champaign, USA, pp. 71−75.

Solomons,G.L. (1969) *Materials and Methods in Fermentation*. Academic Press, London.

Stowell,J.D. and Bateson,J.B. (1983) Economic aspects of industrial fermentations. In *Bioactive Microbial Products. 2. Development and Production*. Nisbet,L.J. and Winstanley,D.J. (eds), Academic Press, London, pp. 117−139.

Swern,D. (1964) *Bailey's Industrial Oil and Fat Products*. Interscience Publishers, a Division of John Wiley and Sons, London.

CHAPTER 9

Choice of substrate in polyhydroxybutyrate synthesis

S.H.COLLINS

*Delta Biotechnology Ltd, Castle Court, Castle Boulevard,
Nottingham NG7 1FD, UK*

Introduction

The economics of substrate choice in single cell protein processes is determined largely by cell yield and substrate cost — these topics have been extensively studied and reviewed (Goldberg *et al.*, 1976; Ratledge, 1977). The economics of the production of simple pure chemicals is much less well defined and the efficiency of conversion of the substrate can be critical. The purpose of this chapter is to demonstrate that if there is an adequate knowledge of the specific biochemistry involved it is possible to understand and even predict the fermentation yields of the product on a particular substrate. The influence of fermentation yield and other factors on the economics of substrate choice are also discussed. Poly-β-hydroxybutyrate (PHB) production has been chosen as an illustrative example, although similar considerations would also apply to other products such as triglycerides or polysaccharides.

PHB is an intracellular storage polymer whose biological function is to provide a reserve of carbon and energy (Dawes and Senior, 1973; Matin *et al.*, 1978). It is made by a range of different bacteria, some members of the groups *Alcaligenes* and *Azotobacter* storing particularly high levels of the polymer. Its production has been studied for several years at ICI as a potentially useful biodegradable thermoplastic (Holmes, 1985).

The metabolic intermediate from which PHB biosynthesis diverges from general metabolism is acetyl CoA. The pathways for PHB synthesis and breakdown to and from this metabolite are given in *Figure 1*. β-Ketothiolase is a key regulatory enzyme in both directions — the cleavage reaction being inhibited by acetoacetyl CoA and stimulated by free CoASH which is also a potent inhibitor of the condensation (Senior and Dawes, 1973; Oeding and Schlegel, 1973). The biosynthetic pathway can be summarized as an equation:

$$2AcCoA + NAD(P)H + PHB_n \rightarrow NAD(P)^+ + 2CoASH + PHB_{n+1} \quad \text{Equation 1}$$

where PHB_n and PHB_{n+1} represent polymer molecules of n and $n+1$ monomer units respectively. Thus the energy required for condensation and polymerization is provided by hydrolysis of the CoA ester links and requires otherwise only the one mole of NAD(P)H to reduce acetoacetyl CoA to hydroxybutyryl CoA.

161

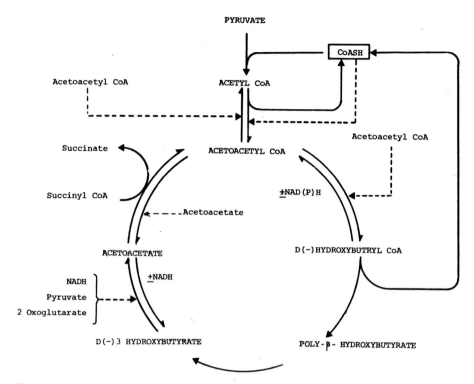

Figure 1. Polyhydroxybutyrate metabolism and regulation.

Choice of substrate

Methanol

In methylotrophs PHB synthesis has been described only in the facultative serine pathway organisms (Anthony, 1982). Acetyl CoA is an intermediate in this pathway which can be drawn as an acetyl CoA synthesizing cycle as in *Figure 2*. In organisms which possess pyruvate dehydrogenase (PDH), acetyl CoA may also be produced from the conventionally drawn pathway via phosphoglycerate (Anthony, 1982). However, the alternative route uses more ATP and is likely to be less efficient. The reactions of *Figure 2* can be summarized with a second equation:

$$2CH_2O + 2NADH + 2ATP + CoASH \rightarrow AcCoA + 2NAD^+ + 2ADP + 2P_i$$
Equation 2

Equations 1 and 2 can be combined to obtain:

$$PHB_n + 2CH_2O + 2CO_2 + 5NAD(P)H + 4ATP \rightarrow$$
$$PHB_{n+1} + 5NAD(P)^+ + 4ADP + 4P_i$$
Equation 3

The CO_2 comes from a two-stage formaldehyde oxidation. If both steps are NAD-linked (Anthony, 1982) this provides 2 mol NAD/mol CO_2 so the total net synthesis from formaldehyde is:

$$PHB_n + 4CH_2O + NAD(P)H + 4ATP \rightarrow PHB_{n+1} + NAD(P)^+ + 4ATP + 4P_i$$
Equation 4

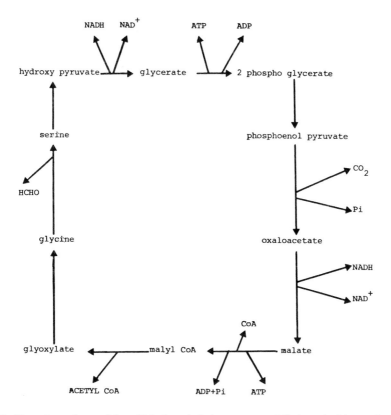

Figure 2. The serine pathway of formaldehyde assimilation as an acetyl CoA synthesizing cycle.

The ATP could be generated by oxidation of about 1.3 mol NADH if a stoicheiometry of 3 mol ATP/mol NADH is assumed or 2 mol NADH if 2 mol ATP/mol is assumed. Thus the total NADH requirement for ATP generation and reduction is about 3 mol requiring 1.5 mol formaldehyde giving a total of 5.5 mol formaldehyde per mol hydroxybutyrate monomer incorporated.

So far only the synthesis from formaldehyde has been considered, but since the real substrate is methanol there should be some allowance for ATP generated by oxidation of methanol to formaldehyde. The methanol dehydrogenase (MDH) − cytochrome oxidase respiratory chain appears to represent a proton-translocating arm (Dawson and Jones, 1982; Jones *et al.*, 1982) and so might provide ATP from the methanol − formaldehyde oxidation with a corresponding reduction in substrate consumption. However, the conditions of methanol concentration required for PHB synthesis may result in excess methanol oxidation. Methanol oxidation in whole cells does not appear to be subject to respiratory control and even transient exposure to excess methanol results in loss of cell yield in SCP production.

Therefore methanol oxidation at high methanol concentrations may not be energy conserving. The difference between the standard redox potentials of the electron donor reaction (methanol/formaldehyde) and the acceptor reduction (oxygen/water) is about 1 volt. Alternatively the proton motive force across a cell membrane is probably unable

163

to exceed about 250 mV (Nicholls, 1974). With only a single net proton translocating event in the respiratory chain, rapid methanol oxidation should still be expected with most of the protons ejected from the cell returning through a non-ohmic leak if the intracellular ATP level is high, assuming a simple chemiosmotic mechanism of respiratory control.

It seems probably best to assume that the methanol−formaldehyde oxidation makes no significant contribution to the energetics of the process and that the methanol requirement for PHB synthesis is effectively the same as for formaldehyde — namely about 5.5 mol/mol hydroxybutyrate or about 2 g methanol per g polymer. The yield of polymer will be further affected by the need to synthesize other biomass so that a high proportion of PHB to cells is required. The highest reported PHB content for a methylotroph is about 67% (Suzuki *et al.*, 1986). Cell yields for serine pathway organisms in carbon limitation have been reported at about 0.37 g cells/g methanol (Goldberg *et al.*, 1976) but values of around half this occur in carbon excess and nitrogen limitation — such as would be used in PHB synthesis. Then 100 g PHB-containing cells would require 134 g methanol to make 67 g PHB and maybe 180 g methanol for the other biomass — a total of 314 g methanol giving an overall yield of about 0.21 g PHB/g methanol. If this is the maximum theoretical yield then little improvement should be expected on the published achieved value of 0.18 g/g methanol (Suzuki *et al.*, 1986) without a substantial improvement in the proportion of PHB to other biomass.

Hydrogen and CO_2

Hydrogen may be considered as an alternative substrate to methanol. Hydrogen oxidising organisms such as *Alcaligenes eutrophus* have been reported to produce levels of PHB approaching 80% of the dry weight (Sonnleitner *et al.*, 1979). Also since ICI uses large quantities of hydrogen as 'synthesis gas' in methanol and ammonia manufacture, the possibility of using hydrogen together with oxygen and carbon dioxide as the substrates for PHB production would be extremely attractive. The energy requirements for the fixation of CO_2 via the Calvin cycle can be summarized in the following equation:

$$6CO_2 + 16ATP + 10NADPH \rightarrow 2PGA + 10NADP^+ + 16ADP + 16P_i$$

Equation 5

If phosphoglyceric acid (PGA) is converted to PHB via phosphoenol pyruvate, pyruvate and acetyl CoA the synthesis reaction becomes:

$$PHB_n + 2PGA + NAD + 2ADP + 2P_i \rightarrow PHB_{n+1} + 2ATP + NADH + 2CO_2$$

Equation 6

So combining 5 and 6 the following is obtained:

$$PHB_n + 4CO_2 + 14ATP + 10NADPH + NAD^+ \rightarrow$$
$$PHB_{n+1} + 14ADP + 14P_i + 10NADP^+ + NADH$$

Equation 7

NADH is produced in *A. eutrophus* growing on hydrogen by reduction of NAD with hydrogenase (Pfitzner *et al.*, 1970) with NADPH production by energy linked trans-hydrogenase (Jones *et al.*, 1975). Substituting NADH and ATP for NADP in Equation 7 the next equation is obtained:

$$4CO_2 + 24ATP + 9NADH + PHB_n \rightarrow PHB_{n+1} + 24ADP + 24P_i + 9NAD^+$$

Equation 8

If it is assumed that one mol NADH can produce 3 mol ATP, 17 mol NADH (or hydrogen) are required to produce the necessary amount of ATP and NADH, giving a minimum hydrogen requirement of 34 g/mol monomer polymerized or 2.5 g polymer/g H_2. However Bongers (1970) concluded, from a study of fermentation yields of *A. eutrophus*, that the yield of ATP may be only 2 mol/mol H_2 which would lower the yield to about 2 g polymer/g hydrogen. Such hypothetical yields for the biosynthetic pathway are comparable with the estimated methanol consumption rates of 2 g/g polymer and must be corrected for hydrogen consumption in non-PHB biomass production under sub-optimal conditions. Fermentation results suggest hydrogen consumption at the rates of around 0.75 g/g cells under such conditions (Bongers, 1970) which would increase the hydrogen requirement by around 40% even if PHB levels aproaching 80% are achieved. In the absence of reliable data on actual PHB yields from hydrogen it might be expected that the achievable yields would be somewhat lower than the theoretical prediction, i.e. probably not better than 1 g PHB/g hydrogen.

Substrate consumption cost is, however, only part of the economic assessment of hydrogen. The flammability, low solubility and large oxygen requirement pose major problems requiring further costs in research, plant construction and materials. Thus low hydrogen solubility requires recycle of exhaust gases to avoid losses. This is impossible unless oxygen is supplied pure, rather than as compressed air (to avoid the need to purge nitrogen) and requires levels of CO in the synthesis gas to be kept low enough to prevent it rising to inhibitory levels. Gas recycle would also require compression of hydrogen/oxygen mixtures which would be difficult to achieve safely. Considerations such as these have led ICI to conclude that hydrogen is an unrealistic substrate for this purpose.

Carbohydrates

Carbohdyrate is potentially a readily accessible substrate for PHB production. Although hydrogen is no longer being considered as a substrate *A. eutrophus* may still be the organism of choice because of its high capacity for PHB storage and its ability to grow on simple sugars. The wild-type organism grows only on fructose, but mutants capable of growth on glucose are well-known (Schlegel and Gottschalk, 1965). However, *A. eutrophus* is incapable of growing on complex carbohydrates such that if sucrose were the substrate of choice it would be necessary to carry out an initial inversion. Consideration of the pathway of PHB synthesis is relatively simple with a source of carbohydrate as the feedstock. Acetyl CoA is produced by the PDH reaction. The energetic equation for the overall reaction for an organism using the Entner-Doudoroff pathway is:

$$PHB_n + Glucose + ADP + P_i + 3NAD^+ \rightarrow PHB_{n+1} + ATP + 3NADH + 2CO_2$$

Equation 9

The pathway is actually energy yielding since it produces ATP and NADH, but because of the loss of CO_2 the yield cannot be better than 1 mol monomer incorporated per mol glucose. Thus the glucose requirement for PHB alone is about 2.1 g/g PHB. If a single stage continuous fermentation is used to produce PHB, energy is also required for biomass production. Maximum PHB yields are only achievable in batch or two-stage processes where growth precedes polymer storage. Allowing for non-PHB biomass on the same basis as before, a minimum glucose requirement of about 2.5 g/g PHB would be predicted — actually observed values in fed batch fermenters at ICI are around 3 g/g PHB.

Organic acids

Clearly substrates which produce acetyl CoA from PDH cannot give better than 67 % carbon conversion to product because of the CO_2 loss in this reaction. Choosing a substrate which is metabolically 'closer' to PHB such that acetyl CoA is not produced from pyruvate or is bypassed altogether might yield greater quantities of PHB with higher yields. Thus crotonate, which requires only hydroxylation and activation to the CoA derivative, is incorporated into PHB with great facility (Gottschalk, 1965). However, with crotonic acid trading at about £2000/tonne this does not seem a realistic choice. Butyric acid might also be converted readily to PHB but is still expensive at around £650/tonne. Acetic acid is the obvious choice as a source of acetyl CoA — some substrate oxidation is required in this case to provide energy for acetate transport, CoA derivatization and NADH production but oxidation of 0.5 mol acetate by the TCA cycle should be adequate for all this, so that the hypothetical acetic acid requirement for PHB synthesis alone would be about 1.75 g/g PHB. Unfortunately acetic acid is a relatively oxidized substrate and is poorly placed for biomass synthesis [i.e. has a low theoretical Y_{max} ATP (Stouthamer, 1973)] with observed yields in carbon limitation around 0.3 g/g substrate. If we allow for acetate requirements arising from biomass production on the same basis as previously, we would predict a minimum requirement of about 2.6 g acetic acid/g polymer. Actual observed values are about 3 g/g polymer.

Ethanol

There is however another more reduced substrate, ethanol, which is readily oxidized to acetic acid and which should give better yields. Wild type *A. eutrophus* does not grow on ethanol (Aragno and Schlegel, 1981) but mutants have been isolated which have this capacity (Senior, Collins and Richardson, 1985). In this case oxidation of ethanol to acetic acid provides more than sufficient NADH for acetoacetyl CoA reduction and ATP for CoA derivatization. Allowing for biomass production predicted requirement of about 1.6 g ethanol/g PHB can be obtained on the same basis as previous calculations. Actually observed values are around 2 g/g polymer.

Economic considerations

It is possible to examine the economics of substrate choice in the light of these expected and achieved substrate yields. *Table 1* lists approximate prices for the substrates discussed. Prices are taken from 'Chemical Marketing Reporter' and 'Financial Times' in early June 1986 and are based upon trading in the USA. Hydrogen is an exception — the price quoted of £500/tonne is calculated on the fuel value which is a minimum estimate but hydrogen may cost up to 100 times that much if transported in tankers or cylinders. The table also lists fermentation yields (estimated only in the case of hydrogen) and in spite of its poor yield, the low cost of methanol makes it apparently the cheapest substrate after hydrogen but it suffers from additional problems of higher product purification cost consequent upon lower PHB content and relatively low molecular weight polymer with inferior properties (Suzuki *et al.*, 1986; Holmes, 1987).

Table 1. Substrate prices, PHB yields and substrate costs for PHB production.

	Approx price (£/tonne)	Yield (g PHB/g substrate)	Substrate costs (£/tonne PHB)
Methanol	110	0.18	610
Ethanol (synthetic)	440	0.5	880
Propanol	600	–	
Acetic acid	370	0.33	1220
Propionic acid	490	–	
Dextrose	360	0.33	1180
Cane sugar (refined)	200	0.33	660
Hydrogen	(500)	(1.0)	(500)

Prices calculated from values published in 'Chemical Marketing Reporter', June 2, 1986 and from the 'Financial Times', June 2, 1986. Hydrogen values estimated as described in text.

Cane sugar seems to be the best carbohydrate source based on *Table 1*, but prices may be misleading because they are based upon refined crystallized products whereas lower grades may be acceptable. Crude sugar costs about £100/tonne and molasses can be anywhere between £0−100/tonne sugar dependent upon quality and location. In the case of PHB production some purification of these crude sugars may be necessary in order to maintain conditions of phosphorus or nitrogen limitation required for polymer production. The price of carbohydrate is also the subject of political control within the EEC so that carbohydrate may be a less attractive carbon source in those member countries.

Ethanol is the best substrate in terms of g product/g C source but at the 'synthetic' ethanol price given in *Table 1*, it seems more costly than carbohydrates. However, fermentation alcohol is available at about 2/3 this value which makes it competitive at least with purified carbohydrates. Ethanol prices are also subject to political factors such as government decisions to build 'gasohol' plants and cheap ethanol might become available in the EEC from agricultural surplus.

Moderate yields and relatively high substrate price appear to rule acetic acid out of consideration. The table also lists two 3-carbon compounds which are of interest in the biopolymer fermentation, but not for pure PHB production. Propanol and propionic acid can produce propanoyl CoA as a metabolic intermediate. β-Ketothiolase catalyses a condensation of this molecule with acetyl CoA with subsequent reduction and polymerization. It has been suggested (Holmes, 1987) that the resulting polymer has better physical properties such as lower melting point and greater toughness. The substrates are relatively expensive and are used with carbohydrates or ethanol as a source of C_2 units. The additional substrate cost of copolymer production depends upon the proportion of hydroxypentanoate required and the efficiency with which the C_3 source is incorporated into the resulting polymer in competition with other fates — such as degradation to acetate and hence possibly the formation of regular PHB.

Conclusion

It is possible that ICI's experience with PHB can be extrapolated to other chemically simple polymers. From this study it would appear that substrates which are metabolically close to the product are likely to give the best yield. Thus for example ethanol would seem to be a promising substrate for triglyceride production whereas carbohydrate would be the obvious choice for a polysaccharide product. Autotrophic systems will always suffer from the disadvantages of the high bioenergetic requirement of CO_2 fixation together with the obvious disadvantages of capital requirement and plant cost incurred. Other C_1 substrates, such as methanol, are liable to suffer from specific problems, such as formaldehyde accumulation in substrate excess. The use of these unconventional substrates can only be justified if their prices are competitive. In the case of PHB fermentation ethanol is as cost-effective as methanol in spite of a 3-fold difference in substrate price.

References

Anthony,C. (1982) *The Biochemistry of Methylotrophs.* Academic Press, London.

Aragno,M. and Schlegel,H.G. (1981) The hydrogen-oxidizing bacteria. In *The Procaryotes.* Starr,M.P., Stolp,H.G., Trouper,H.G., Balows,A. and Schlegel,H.G. (eds), Springer-Verlag, Berlin, pp. 865−893.

Bongers,L. (1970) Energy generation and utilization in hydrogen bacteria. *J. Bacteriol.,* **104,** 145−151.

Dawes,E.A. and Senior,P.J. (1973) The role and regulation of energy reserve polymers in micro-organisms. *Adv. Microb. Physiol.,* **10,** 135−266.

Dawson,M.J. and Jones,C.W. (1982) Respiration-linked proton translocation in the obligate methylotroph *Methylophilus methylotrophus. Biochem. J.,* **194,** 915−924.

Goldberg,I., Rock,J.S., Ben-Bassat,A. and Mateles,R.I. (1976) Bacterial yields on methanol, methylamine, formaldehyde and formate. *Biotechnol. Bioeng.,* **18,** 1657−1668.

Gottschalk,G. (1965) Die verwertung organischer substrate durch *Hydrogenomonas* in gegenwart von molekularem wasserstoff. *Biochem. Z.,* **341,** 260−270.

Holmes,P.A. (1985) Applications of PHB − a microbially produced biodegradable thermoplastic. *Phys. Technol.,* **16,** 32−36.

Holmes,P.A. (1987) Biologically produced 3-hydroxyalkanoate polymers and copolymers. In *Developments in Crystalline Polymers, Vol. 2.* Bassett,D.C. (ed.), Applied Science Publishers, London, in press.

Holmes,P.A., Wright,F.W. and Collins,S.H. (1980) European Patent Specification 0052459.

Jones,C.W., Brice,J.M., Downs,A.J. and Drozd,J.W. (1975) Bacterial respiration-linked proton translocation and its relationship to respiratory-chain composition. *Eur. J. Biochem.,* **52,** 265−271.

Jones,C.W., Kingsbury,S.A. and Dawson,M.J. (1982) The partial resolution and dye-mediated reconstitution of methanol oxidase activity in *Methylophilus methylotrophus. FEMS Microbiol. Lett.,* **13,** 195−200.

Matin,A., Veldhuis,C., Stegeman,V. and Veenhuis,S. (1979) Selective advantage of a *Spirillum* sp. in a carbon limited environment. Accumulation of poly-hydroxybutyric acid and its role in starvation. *J. Gen. Microbiol.,* **112,** 349−355.

Nicholls,D.G. (1974) The influence of respiration and ATP hydrolysis on the proton-electrochemical gradient across the inner membrane of rat-liver mitochondria as determined by ion distribution. *Eur. J. Biochem.,* **50,** 305−315.

Oeding,V. and Schlegel,H.G. (1973) β-Ketothiolase from *Hydrogenomonas eutropha* H16 and its significance in the regulation of poly-hydroxybutyrate metabolism. *Biochem. J.,* **134,** 239−248.

Pfitzner,J., Linke,H.A.B. and Schlegel,H.G. (1970) Eigenschaften der NAD-spezifischen Hydrogenase aus *Hydrogenomonas* H16. *Arch. Microbiol.,* **71,** 67−78.

Ratledge,C. (1977) Fermentation substrates. *Annu. Rep. Ferment. Processes,* **1,** 49−72.

Schlegel,H.G. and Gottschalk,G. (1985) Verwertung von glucose durch eine mutante von *Hydrogenomonas* H16. *Biochem. Z.,* **341,** 249−259.

Senior,P.J., Collins,S.H. and Richardson,K.R. (1985) British Patent application 204442.82.

Senior,P.J. and Dawes,E.A. (1973) The regulation of poly-hydroxybutyrate metabolism in *Azotobacter beijerinckii. Biochem. J.,* **134,** 225−238.

Sonnleitner,B., Heinzle,E., Braunnegg,G. and Lafferty,R.M. (1979) Formal kinetics of poly-hydroxybutyric acid production in *Alcaligenes eutrophus* H 16 and *Mycoplana rubra* R 14 with respect to the dissolved oxygen tension in ammonium-limited batch cultures. *Eur. J. Appl. Microbiol. Biotechnol.,* **7,** 1−10.

Stouthamer,A.H. (1973) A theoretical study on the amount of ATP required for synthesis of microbial cell material. *Antonie van Leeuwenhoek,* **39,** 545−565.

Stouthamer,A.H. (1972) Influence of hydrogen acceptors on growth and energy production of *Proteus mirabilis. Antonie van Leeuwenhoek,* **38,** 81−90.

Suzuki,T., Yamano,T. and Shimizu,S. (1986) Mass production of poly-hydroxybutyric acid by fully automatic fed-batch culture of methylotroph. *Appl. Microbiol. Biotechnol.,* **23,** 322−329.

CHAPTER 10

The link between the efficiency of ATP generation and metabolic flexibility during anaerobic growth on various carbon sources

M.J.TEIXEIRA DE MATTOS[1], H.STREEKSTRA[1], O.M.NEIJSSEL[1] and D.W.TEMPEST[2]

[1]*Laboratory for Microbiology, University of Amsterdam, Amsterdam, The Netherlands, and* [2]*Department of Microbiology, University of Sheffield, Sheffield, UK*

Introduction

In heterotrophic organisms a single carbon substrate may serve to provide the cell with both the intermediary metabolites needed for biosynthesis and the energy required for this process. Thus, after being transported into the cell, metabolism of the (activated) carbon substrate serves to generate two flows; that is, a catabolic and an anabolic one. In addition, because catabolism is an oxidative process, it is always accompanied by a flow of reductant. With carbon substrates that are more reduced than biomass, this is also true for anabolism. In *Figure 1*, the three flows are schematically represented. For balanced growth to proceed, it is a prerequisite that the above-mentioned flows of intermediary metabolites, reductant and energy (ATP) are closely matched in such a manner that all reducing equivalents are re-oxidized and all of the energy generated is dissipated, either in anabolism or by energy-dissipating mechanisms not associated with growth. So, the rate of catabolism is related to the rate of anabolism both by the energetic demands needed to sustain the biosynthesis and by the occurrence of energy-spilling reactions. As for the flow of reductant, this is coupled to energy generation either directly, under conditions of respiratory activity, or indirectly under conditions in which energy is generated by substrate level phosphorylation only.

Under conditions that allow respiration, the direct coupling of the re-oxidation of reducing equivalents to energy generation implies that the partitioning of substrate between catabolism and anabolism can be governed solely by the energetic demand of biomass formation under a particular growth condition. Any shortfall in energy can be circumvented by directing more substrate towards catabolism. However, with many organisms a more complex metabolic response to the energetic demands provoked by the environment has been observed (Stouthamer and Bettenhausen, 1975; Neijssel and Tempest, 1975; Hommes *et al.*, 1985). Thus, aerobically grown cultures of *Klebsiella*

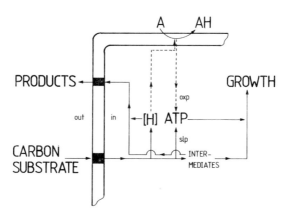

Figure 1. Schematic presentation of the carbon, energy and reductant flow in a cell. slp: substrate level phosphorylation; oxp: oxidative phosphorylation; A/AH: oxidized/reduced electron acceptor; [H]: reduced redox carrier; :- - - - - with an external electron acceptor present.

aerogenes react to conditions in which the energy source is present in excess by excreting a number of products that are usually more oxidized than the substrate (Neijssel and Tempest, 1975, 1979). This phenomenon has been termed overflow metabolism. Because production of overflow metabolites represents a flow to partially oxidized products, this flow will cause a decreased efficiency in the use of the carbon substrate. In this sense, overflow metabolism constitutes a mechanism to vary the efficiency of energy generation; that is, a mechanism to partially uncouple the flows of catabolism and anabolism (Neijssel and Tempest, 1976). For this reason one can define overflow metabolism as a metabolic reaction that results in a departure from the theoretically most efficient use of the energy source.

When an organism is growing in the absence of an external electron acceptor, three additional complications arise in balancing the anabolic, catabolic and reductant flows. First, energy can be generated only by substrate level phosphorylation. As a consequence, the stoicheiometry of energy generation (mol ATP produced per mol substrate consumed) is low and therefore the rate of catabolism has to be high. Second, in contrast to oxidative phosphorylation — where mechanisms of by-passing phosphorylation sites (Neijssel *et al.*, 1978) or modification of the respiratory chain (Jones, 1977) may affect P/O ratios — substrate level phosphorylation reactions are strictly coupled processes and therefore the variability in the stoicheiometry of ATP generation should be significantly diminished under fermentative conditions. Third, energy generation is generally accompanied by formation of reductant. To balance this flow, carbon substrate is needed to provide intermediates that can serve as electron acceptors.

Despite these restrictions in flexibility, many organisms — including the so-called homofermentative streptococci (Thomas *et al.*, 1979) — are able to ferment carbon substrates to a variety of end products and the spectrum of products formed depends strongly upon the growth conditions (Gale and Epps, 1942; Blackwood *et al.*, 1956; Wallace, 1978; Thomas *et al.*, 1979; Tran Din and Gottschalk, 1985). By analogy with aerobic cultures, the concept of overflow metabolism can also be applied to fermenting cultures. For fermenting cells, the maximal energetic efficiency is determined by the biochemical make up of the microbial species under study. Nowadays the biochemistry of many fermentation pathways is known, allowing accurate calculations to be made of the rate of ATP synthesis during catabolism (Gottschalk and Andreesen, 1979; Thauer *et al.*, 1977).

For the reasons mentioned above, it can be concluded that under fermentative conditions the interrelationships between the catabolic, anabolic and reductant flow are rather complex. The question then arises as to how organisms cope with environmental conditions where a disturbance of the various flows is to be expected. Since we will discuss the metabolic strategies concerning the rate of energy generation, the transition of a growth environment limited by the energy source to one in which the energy source is present in excess, is of special interest. Here, two responses seem possible. It may be that the consumption of the substrate is strictly tuned to the energetic demands provoked by that particular growth environment and no instantaneous reaction will be observed [for example, see Kornberg (1973)]: catabolism and anabolism are fully coupled and any change in the ratio of the two flows will lead to an imbalance of the ATP/ADP ratio. Obviously, in this case the organisms are extremely unreactive with respect to changes in their growth environment. Only when the organisms are able to increase their growth rate instantaneously, will an increase in the overall carbon substrate metabolism be observed. The other possibility is the presence of uncoupling mechanisms that allow the organism to lower the stoicheiometry of energy generation. Partial uncoupling may be brought about either by invoking a shift in the fermentation pattern with a decreased energetic efficiency or completely by invoking an active ATP-hydrolysing system. Such mechanisms would allow a direct reaction to environmental changes and endow the organism with metabolic flexibility. The advantage of such flexibility is obvious, bearing in mind that in many natural environments steady-state conditions are rare and the potential to adapt rapidly to a new condition is of paramount importance. It is beyond the scope of this chapter to discuss the ecophysiological significance of metabolic flexibility (or, as it is often called, metabolic reactivity) (see Koch, 1971; Tempest and Neijssel, 1978; Harder and Dijkhuizen, 1983; Tempest *et al.*, 1983). Here, it is important to stress that the extent to which an organism expresses flexibility will affect the usefulness of its application in, say, the production of chemicals. Therefore, in this chapter we shall focus on the effect of the nature of the carbon substrate on the various catabolic flows and on metabolic flexibility.

The study of the bioenergetics of microbial growth

Early studies on the energetics of microbial growth showed a strong correlation between the amount of biomass formed and the amount of energy (ATP) generated during catabolism (Bauchop and Elsden, 1960; Rosenberger and Elsden, 1960; Beck and Stugart, 1966; Brown and Collins, 1977). Calculations of the yield of biomass per mol ATP synthesized (Y_{ATP}) relied upon knowledge of the catabolic pathways and the stoicheiometry of ATP synthesis. The stoicheiometry of oxidative phosphorylation is not only difficult to assess but may vary with varying growth conditions (Jones, 1977). For this reason, growth conditions in which ATP synthesis occurs exclusively by substrate level phosphorylation are better suited for studying the energetics of growth. Here, the coupling between ATP synthesis and the formation of fermentation products is assumed to be unequivocal. Thus, once the specific rates of end product synthesis are determined, the specific rate of ATP synthesis is known. Since in batch cultures the fermentation pattern (and hence the stoicheiometry of ATP synthesis) may vary throughout the growth cycle, such cultures are less suited to quantitative studies on the bioenergetics of growth. In contrast, the growth of organisms under steady-state conditions allows precise calculations of energy flows. The energetic demands of cell synthesis in a particular growth environment can then be quantified. Chemostat cultures are most suitable for these studies.

In this chapter, we concentrate on the metabolic behaviour of *K. aerogenes* during steady-state and transient state growth conditions. This organism was grown anaerobically in chemostat culture on carbon sources of varying degrees of reduction, yielding varying amounts of ATP per mole of substrate fermented. By expressing both the rate of production formation and biosynthesis as specific rates (q, mmol/g cells^{-1}/h^{-1}), the specific ATP generation rate could then be calculated for steady-state conditions. As has been pointed out previously, during conditions of steady-state growth:

(i)   the net flow of reductant must be zero, therefore $NADH_2$ consumption and production rates must be equal; and

(ii)   the net flow of ATP must be zero, therefore the sum of the rate of ATP consumed by biosynthesis and hydrolysed by mechanisms not associated with growth must equal the rate of ATP synthesis during product formation.

Thus, the energetic demands of biosynthesis during steady-state growth (Y_{ATP}) can be calculated by determining the catabolic flows. It is important to note that the assessment of Y_{ATP} is only valid if all carbon consumed can be accounted for as products and biomass:

$$q_C\text{-consumed} = q_C\text{-biomass} + q_C\text{-product } 1 + q_C\text{-product } 2 + \ldots$$
$$+ \ldots + q_C\text{-product } n$$

In our calculations, we assumed the overall cell composition to be $C_4H_7O_2N$ (Herbert *et al.*, 1971). Calculations on Y_{ATP} and the reductant flow were based on the assumption that anabolism starts at the level of pyruvate. Thus, with growth on, say, glucose, ATP is synthesized during the formation of the pyruvate needed for biosynthesis. With this method, the energy required for gluconeogenesis from pyruvate is taken to be equal to the energy gain from the dissimilation of C_6 compounds to pyruvate but this inaccuracy does not significantly affect the calculations (see Stouthamer, 1973).

With the data obtained with steady-state cultures, metabolic responses to changes in the environment can be quantified. The only assumption made is that the energetic demands of biosynthesis are equal under steady state and transient state conditions. By adding a saturating amount of the carbon substrate to a carbon-limited chemostat culture, and making the same calculations as for steady-state cultures, both the capacity of anabolism and catabolism as well as the extent of uncoupling of catabolism and energy generation can be assessed. The following results show that metabolic flexibility is strongly dependent on the potential to uncouple the energy flow from catabolism.

Steady state growth

In the catabolism of glucose, gluconate or mannitol, phosphoenol pyruvate (PEP) is a central intermediary metabolite (see *Figure 2*). From here, a branching of the catabolic

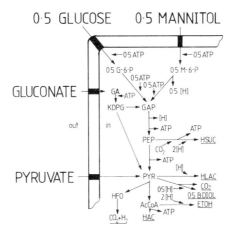

Figure 2. Fermentation pathways of glucose, mannitol, gluconate and pyruvate by *K. aerogenes*. G-6-P: glucose-6-phosphate; M-6-P: mannitol-6-phosphate; GA: gluconate; KDPG: 2-keto-3-deoxy-6-phosphogluconate; GAP: glyceraldehyde-3-phosphate; PEP: phosphoenol pyruvate; HSUC: succinic acid; PYR: pyruvate; HLAC: lactic acid; AcCoA: acetyl coenzyme A; B.DIOL: 2,3-butanediol; HFO: formic acid; HAC: acetic acid; EtOH: ethanol; [H]: $NADH_2$.

flow may occur resulting in the formation of pyruvate or succinate. Pyruvate may be converted subsequently to acetyl-CoA, 2,3-butanediol or D-lactate. Finally, acetyl-CoA can lead to the synthesis of acetate or ethanol, with the concomitant formation of formate and/or CO_2 and H_2. With the exception of acetate formation, synthesis of these end products is accompanied by the generation of 1 mol of ATP. Acetate synthesis, however, is energetically more efficient because one extra mole of ATP is generated in the conversion of acetyl phosphate to acetate. With this scheme, the maximal amount of ATP generated per mole of substrate fermented can be calculated. Taking into account the need of redox neutrality, it can be concluded that, with glucose as the carbon substrate, production of equimolar amounts of ethanol (or succinate) and acetate constitutes the maximal energetic efficiency possible under anaerobic conditions (3 mol of ATP per mol of substrate fermented). This fermentation pattern was observed under glucose-limited conditions only (see *Table 1*). Under glucose-excess conditions the energetic efficiency was always lower. Yet, this loss in efficiency was more than compensated for by an increase in the total catabolic flow resulting in an increased rate of energy generation. Thus, under glucose-excess conditions, the Y_{ATP} value was lowered. It can be argued that extra energy is needed to drive the uptake and assimilation of the respective growth-limiting nutrients from the environment. The uptake of the limiting substrate occurs often via energetically more expensive transport systems [Meers *et al.*, 1970; Mulder *et al.*, 1986; see also Tempest and Neijssel (1984)]. The point we wish to make here, however, is that the shift in the fermentation pattern towards a less efficient ATP-generating catabolic flow may be regarded as a mechanism to uncouple catabolism from anabolism. According to the definition given above, here lactate and 2,3-butanediol may thus be considered as overflow metabolites. In this respect, anaerobic cultures behaved similarly to aerobic cultures (Neijssel and Tempest, 1975, 1976): under energy excess conditions, the metabolic strategy is directed towards a variable efficiency of ATP generation and high catabolic flows.

173

Table 1. Rates of glucose utilization and of product formation expressed in variously-limited chemostat cultures of K. aerogenes growing at D = 0.4/h (pH 6.8, 35°C).

	Growth limitation				
	Glucose	Phosphate	Ammonia	Sulphate	Potassium
Glucose used	85.0	111.8	123.1	144.1	164.4
CO_2 + formate	19.6	27.2	29.7	34	37.4
Acetate	23.1	21.0	23.1	26.1	34.8
Ethanol	20.1	23.5	25.7	29.5	34.6
D-Lactate	0.3	4.5	3.1	7.5	8.3
2,3-Butanediol	0	10.1	11.2	12.3	6.2
Succinate	0	2.7	3.6	3.7	4.1
$NADH_2/NAD \times 100$	87	86	97	98	97
Carbon recovery	98	101	95	94	89
$Y_{glucose}$	35	27	24	21	18
Y_{ATP}	12.5	10.5	9.9	8.6	7.6
Mol ATP per mol glucose fermented	3.1	2.6	2.6	2.5	2.4

Values are expressed as milliatom carbon/h, normalized to a cell production rate of 20 milliatom carbon/h. Data from Teixeira de Mattos and Tempest (1983).

Table 2. Specific production rates and yield values of various carbon-limited chemostat cultures of K. aerogenes (D = 0.20 ± 0.01/h, pH 6.9, 35°C).

	Carbon substrate:				
	Glucose	Mannitol	Gluconate	Pyruvate	Glycerol
q (mmol/g cells/h)					
Acetate	6.1	3.7	15	31.1	5.6
Ethanol	6.1	10.0	4	0	11.0
CO_2 + formate	12.2	12.7	18.5	33.3	15.4
D-Lactate	0	0	0	0	0
Succinate	0	1.0	0.5	1.1	1.2
2,3-Butanediol	0	0	0	0	0
1,3-Propanediol	0	0	0	0	13.0
$NADH/NAD \times 100$	100	88	93	?	108
Carbon recovery (%)	100	96	103	97	102
$Y_{substrate}$	27	22	18.5	5.4	6.4
Y_{ATP}	9.5	9.5	9.8[a]/7.3[b]	6.4	8.0
Mol ATP per mol substrate fermented	3.0	2.4	2.6	0.9	0.78

[a]Assuming that uptake of 1 mol of gluconate consumes 0.5 mol of ATP.
[b]Assuming that gluconate uptake does not consume energy.
$Y_{substrate}$ in g cells/mol substrate consumed^{-1}; Y_{ATP} in g cells/mol ATP synthesized.
Values for glucose-limited cultures are obtained by interpolation of data from Teixeira de Mattos and Tempest (1983). Other data from Streekstra et al. (1987a,b).

With a substrate more reduced than glucose, the fraction of substrate that can be catabolized to acetate must be lower. As a consequence, the maximal amount of ATP generated per mole of substrate consumed during fermentation will be less and this will be reflected in the value of $Y_{substrate}$. This was indeed found to be the case with growth on mannitol (*Table 2*). Here, the maximal energetic efficiency of fermentation can be only 2.5 mol ATP per mol substrate fermented, due to the fact that acetate production must be accompanied by ethanol and/or succinate formation in a ratio of 3 to 1 in order to balance the flow of reductant. However, biomass formation from mannitol also generates $NADH_2$. To compensate this flow, an increased catabolic flow to reduced end products (for example ethanol) is needed and so the efficiency of ATP generation cannot exceed 2.4 mol ATP per mol of mannitol fermented. This efficiency was indeed observed during mannitol-limited growth and in this respect the organisms behaved the same as glucose-limited cells. As expected, the Y_{ATP} values for glucose- and mannitol-grown cells were similar.

K. aerogenes ferments gluconate via the Entner−Doudoroff pathway (*Figure 2*). The key intermediate in this route, 2-keto-3-deoxy-6-phosphogluconate, is converted to dihydroxyacetone-phosphate and pyruvate. So, formation of 2 mol of pyruvate from gluconate yields 1 mol of ATP plus one pair of reducing equivalents. Here, biomass formation is a reductive process, therefore the maximum efficiency of fermentation can be slightly over 2.5 mol of ATP per mol of gluconate fermented. Again, carbon-limited cultures catabolized the substrate with this maximal efficiency (*Table 2*). How-ever, it was observed that at all growth rates (Streekstra *et al.*, 1987b) the Y_{ATP} value was 20% lower than with growth on glucose or mannitol. Since there is no reason to assume a difference in the energetic demands of cell synthesis from either of these two substrates, re-calculation of the net energy generation rate during catabolism on the basis of a Y_{ATP} equal to Y_{ATP} for growth on glucose, led to the conclusion that net on-ly 2.1 mol of ATP was formed per mol of gluconate fermented. We suggest that the loss of 0.5 mol of ATP is due to the necessity to invest (the equivalent of) 0.5 mol of ATP for the transport of 1 mol of gluconate into the cell. In this connection, it should be mentioned that it has been shown that in *Escherichia coli* gluconate is transported via a proton symport system (Ramos and Kaback, 1977). It seems logical then that ATP is hydrolysed by an ATPase in order to translocate the protons necessary for this symport. Interestingly, similar results have been reported for the transport of glucose via the galactose−proton symport system in *Salmonella typhimurium* (M.Driessen *et al.*, 1987) and for lactose transport in *E. coli* (Muir *et al.*, 1985).

With the above-mentioned substrates, branched fermentation arises from the central metabolite, PEP. One might ask the question how metabolic flexibility is affected by a substrate that enters into catabolism beyond this intermediate. For this purpose, *K. aerogenes* was grown on pyruvate. Here, ATP only can be generated via acetate for-mation (*Figure 2*). This reaction sequence is redox neutral because under anaerobic conditions decarboxylation of pyruvate is believed to be catalysed by the pyruvate−formate lyase system. Moreover, it has been suggested that the pyruvate dehydrogenase (PDH) complex cannot function anaerobically (Hansen and Henning, 1966). This would be due to the high NADH/NAD ratio usually present in anaerobically growing cells. However, this may not be true for growth on pyruvate and the presence of an active PDH complex cannot be excluded. Another possibility to generate reductant needed for anabolism is, of course, via the glyoxalate cycle. This would explain the produc-tion of succinate by pyruvate-limited cultures (*Table 2*).

Reasoning along similar lines as for gluconate-limited cells, the low apparent Y_{ATP} value for growth on pyruvate suggests that the net ATP synthesis rate is highly over-

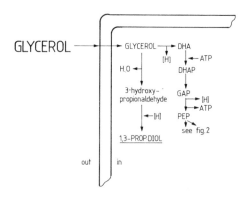

GLYCEROL

Figure 3. Fermentation pathway of glycerol by *K. aerogenes*. DHA: dihydroxacetone; DHAP: dihydroxy-acetone phosphate; 1,3-PROP.DIOL: 1,3-propanediol; GAP, PEP, [H]: see *Figure 2*.

estimated. To our knowledge, no data are available on the mechanism of pyruvate uptake but our results could indicate that energy is needed for the transport of this substrate.

Glucose, gluconate, mannitol and, possibly, pyruvate have in common that they are actively transported into the cell, be it via different mechanisms. Generally, such transport mechanisms have a high affinity for the substrate. In contrast to this, glycerol enters the cell by facilitated diffusion. Under anaerobic conditions, glycerol is either oxidized by glycerol dehydrogenase to yield dihydroxyacetone or dehydrated to 3-hydroxypropionaldehyde (Forage and Foster, 1982; Lin, 1976). The latter intermediate is then reduced to 1,3-propanediol whilst the former enters glycolysis after phosphorylation (*Figure 3*). 1,3-Propanediol formation is necessary in order to balance the generation of $NADH_2$ during biosynthesis. Thus, the maximal efficiency of fermentation (when the remaining glycerol is completely catabolized to ethanol) should be slightly under 1 mol of ATP per mol of glycerol fermented. However, this efficiency was not observed under glycerol-limited conditions: at a growth rate of 0.2/h about 50% of the glycerol was fermented to 1,3-propanediol. These results are in sharp contrast to those of other carbon-limited cultures and it was surprising that under glycerol-limited conditions such high activities of the 1,3-propanediol synthesizing system were present. We suggest, however, that this behaviour reflects a strategy to optimize the *rate* of ATP generation rather than the *efficiency* of energy generation. This view agrees with the conclusions of Westerhoff *et al.* (1983): by an analysis based upon non-equilibrium thermodynamics these workers showed that the thermodynamic efficiency of microbial metabolism is low but, as a consequence, optimal for high anabolic rates. The essential difference between the initial step in the dissimilation of glycerol and the other substrates is that the former is NAD dependent whilst the latter is ATP/PEP dependent. Therefore, it seems logical that the scavenging capacity for glycerol is increased by a mechanism that re-oxidizes at a high rate the reductant initially formed in glycerol oxidation, especially bearing in mind the fact that glycerol dehydrogenase has a low affinity for glycerol [the apparent K_m of the dehydrogenase for glycerol is ~20 mM (Lin, 1976)]. If the assumption is valid that metabolism is directed towards an optimization of rates, it is expected that at higher growth rates — where higher rates of ATP synthesis are needed to sustain anabolism — an increasing fraction of glycerol will be funnelled in the catabolic

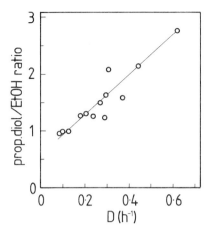

Figure 4. The relationship between the 1,3-propanediol/ethanol production ratio and the growth rate of anaerobic, glycerol-limited cultures of *K. aerogenes* (pH 6.9, 35°C). Data from Streekstra *et al.* (1987a).

flow to 1,3-propanediol. As can be seen from *Figure 4* this is indeed the case. Thus, although a considerable amount of the available energy source is not used as such, the flux of carbon into glycolysis is vastly increased.

Transient state response

From the steady-state data presented above, it can be concluded that under carbon-limited growth conditions the energy source generally is catabolized with a maximal efficiency of ATP generation. With the exception of growth on pyruvate, carbon-excess conditions led to a decrease in efficiency due to the production of overflow metabolites (for glucose see *Table 1*; other data from Streekstra *et al.*, 1987b). It was observed that even for glycerol-grown cells, the efficiency — though not at its theoretical maximum — was always higher during glycerol-limited growth than during glycerol-excess (Streekstra *et al.*, 1987a). We conclude that the significance of branched fermentation resides in the potential to uncouple catabolism from energy generation. However, this potential is considerably less than in the case of aerobic growth and the question arises as to how anaerobic cells, growing carbon-limited, respond to transient states where the supply of the energy source is no longer restricted.

When a glucose-limited culture was relieved of its limitation by a sudden addition of a saturating 'pulse' of glucose, the metabolic rate was increased instantaneously at all growth rates (Teixeira de Mattos *et al.*, 1984) (*Figure 5*). Since at lower growth rates no immediate increase in the rate of cell synthesis was observed, this simply reflected an overcapacity in catabolism. Moreover, a dramatic shift was observed in the fermentation pattern (Teixeira de Mattos and Tempest, 1983) (*Figure 6*): 50% of the extra glucose fermented was catabolized to D-lactate, whilst the remainder was recovered as 2,3-butanediol and succinate. No increase in the acetate production rate was observed and therefore it follows that the increase in the rate of catabolism was not accompanied by a strictly proportional increase in the rate of ATP synthesis. Nevertheless, calculations according to *Figure 2* suggest that the rate of energy generation must have been increased. However, it was found (Teixeira de Mattos *et al.*, 1984)

177

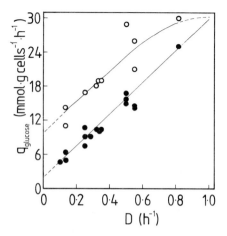

Figure 5. The relationship between the growth rate and the specific rate of glucose consumption (q) of glucose-limited cultures of *K. aerogenes* (pH 6.8, 35°C). (●) specific uptake rate at steady state; (○) specific uptake rate after addition of 10 mM glucose to the culture. Data from Teixeira de Mattos *et al.* (1984).

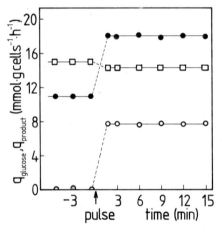

Figure 6. Typical change in the specific glucose uptake and fermentation rates of *K. aerogenes* after relief of glucose limitation ($D = 0.35$ h^{-1}, pH 6.8, 35°C). (●) 1 glucose, (○) $q_{D\text{-lactate}}$, (□) $q_{acetate} + q_{ethanol}$. Data from Teixeira de Mattos and Tempest (1983).

that glucose-limited cells possessed the enzymes of the so-called methylglyoxal by-pass (Cooper and Anderson, 1970; Hopper and Cooper, 1971; Cooper, 1984) (*Figure 7*). It should be noted that D-lactate synthesis from glucose via this by-pass is an ATP-consuming reaction sequence. Moreover, calculations based on the assumption that all of the D-lactate formed resulted from the operation of this by-pass, suggest that the additional catabolic flow was almost completely uncoupled from ATP synthesis (*Table 3*) at growth rates below 0.5/h. Of significance in this connection was the observation that bacteria growing at higher rates were able to increase their growth rate immediately; with these cultures proportionally less of the extra glucose was fermented to D-lactate. Thus, under these latter conditions the efficiency of ATP formation remained high in parallel with the increased energetic demands of biosynthesis. In addition, the newly

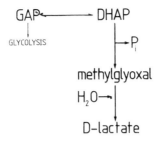

Figure 7. The methylglyoxal by-pass. GAP: glyceraldehyde-3-phosphate; DHAP: dihydroxyacetone phosphate; P_i: inorganic phosphate.

Table 3. Effect of relief of glucose-limited conditions on specific uptake rate (third column), specific ATP synthesis rate (5th and 6th columns) and growth rate (7th column) following various steady-state growth rates.

D (per h)	$q_{glucose}$		q_{ATP}			Specific growth rate after pulse (per h)	Efficiency[c]
	Steady-state	After pulse	Steady-state	After pulse			
				1^a	2^b		
0.13	6.4	11.3	18.3	20.2	30.2	0.13	0.4
0.25	9.4	17.5	26.5	26.8	47.8	0.25	0.3
0.35	10.3	18.2	29.1	31.7	49.3	0.35	0.3
0.51	17.3	29.0	48.3	60.5	77.3	0.60	1.0
0.55	15.7	24.8	43.3	61.0	65.0	0.70	1.9
0.82	25.0	30.5	69.3	83.0	83.0	1.15	2.5

Rates are expressed in nmol/g cells/h.
[a] q_{ATP} when it is assumed that all D-lactate is formed via methylglyoxal.
[b] q_{ATP} when it is assumed that all D-lactate is formed glycolytically.
[c] Mol of ATP per mol of extra glucose fermented, assuming that all D-lactate is formed via the methylglyoxal by-pass.
Data from Teixeira de Mattos *et al.* (1984).

attained growth rate after the pulse was within reasonable limits compatible with the newly attained energy generation rate, as could be deduced from ATP synthesis rates at various steady-state growth rates (data not shown). This suggested that all of the extra ATP formed was turned over by anabolic processes.

It can be concluded that the methylglyoxal by-pass is an exceedingly effective mechanism for uncoupling ATP synthesis from catabolism. A reaction sequence that serves to uncouple energy generation and catabolism can function effectively anaerobically only if (i) it consumes ATP and (ii) redox neutrality can be maintained. With glucose as the energy source, the by-pass itself is redox neutral. It was therefore interesting to investigate whether the methylglyoxal by-pass would function with substrates with degrees of reduction different from glucose.

Although gluconate is catabolized via another route, its fermentation involves the formation of dihydroxyacetone phosphate. This allows, at least theoretically, the use of the methylglyoxal by-pass. Therefore, we investigated the metabolic behaviour of gluconate-limited cells after relief of the limitation (*Figure 8*). Apart from a difference in the products excreted (due to a different degree of reduction of the substrate), gluconate-grown bacteria behaved similarly to glucose-grown bacteria: here, D-lactate and

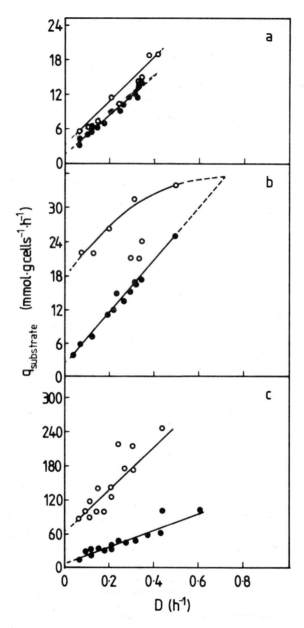

Figure 8. The relationship between the growth rate and the specific rate of substrate consumption (q) of mannitol-**(a)**, gluconate-**(b)** and glycerol-**(c)** limited cultures of *K. aerogenes* (pH 6.8, 35°C). (●) specific uptake rate at steady-state; (○) specific uptake rate after addition of 10 mM of the limiting carbon source to the culture. Data from Streekstra *et al.* (1987a,b).

pyruvate appeared as new products after relief of the limitation and the overall rate of catabolism was increased instantaneously without a concomitant increase in the anabolic flow. Again, the presence of the methylglyoxal by-pass as detected *in vitro* (Streekstra *et al.*, 1987b) would allow (part of) D-lactate production to occur via this energy-dissipating system. In contrast to glucose-grown bacteria, however, it could be calculated that only about 25% of all D-lactate formed must have been synthesized via this by-pass if the rate of ATP synthesis was maintained at a value equal to that under steady-state conditions. The rates of formation of D-lactate by reduction of pyruvate and by the by-pass, respectively, cannot be assessed. However, the excretion of pyruvate after relief of the limitation indicates the presence of a high intracellular pool of pyruvate. Furthermore, since it seems logical to assume that the NADH pool (or redox balance) governs the rate of D-lactate synthesis from pyruvate and that the ATP pool (or energy balance) governs the partitioning of catabolism between glycolysis and the methylglyoxal by-pass, it may be anticipated that, after relief of the limitation, D-lactate is formed through both glycolysis and the methylglyoxal by-pass.

With mannitol as carbon source, the organisms were found not to be able to invoke significant changes in anabolism or catabolism (*Figure 8a*). Interestingly, the activity of the methylglyoxal by-pass was only about 50% of that detected with glucose-grown bacteria (Streekstra *et al.*, 1987b). In this connection, it should be noted that lactate mation from mannitol is an oxidative process. For the methylglyoxal by-pass to be functional, it is therefore necessary that it is accompanied by an overcapacity of reductive catabolic flows. It may well be that this additional prerequisite renders the methylglyoxal by-pass ineffective with reduced substrates.

From *Figure 2* it is clear that, with pyruvate as the energy source, the catabolic flow is severely restricted to a linear fermentation. Moreover, dihydroxyacetone phosphate (the substrate for methylglyoxal synthase) is not an intermediary metabolite in pyruvate catabolism. Upon relief of pyruvate-limited growth, almost no flexibility was observed in the overall metabolism. Only a small increase in the catabolic flow was found without any shift in the fermentation pattern, and anabolic overcapacity was not observed at all. From the results with mannitol- or pyruvate-grown bacteria, we conclude that the absence of a functional uncoupling mechanism resulted in a loss of metabolic flexibility, due to the limited potential to uncouple catabolic and energy flows.

Similar to glucose-limited conditions, *K. aerogenes* was able to increase the specific uptake rate of glycerol after relief of the limitation (*Figure 8c*). However, with glycerol-grown bacteria only a shift in the proportions of the fermentation products was observed; although methylglyoxal by-pass activity could be measured *in vitro* (Streekstra *et al.*, 1987a), no products that possibly were synthesized via this pathway — like 1,2-propanediol (Tran Din and Gottschalk, 1985) — could be detected. Moreover, it was observed that the enhanced catabolic activity was often accompanied by a shut-down of anabolism. Thus, it seems that relief of glycerol limitation resulted in a very significant increase in the ATP synthesis rate. The question then arises as to how this surplus of ATP is dissipated. It can be proposed that the excretion of 1,3-propanediol is an energy-consuming process. The observation of McGee and Richards (1981) that glycerol dehydratase is membrane-associated is compatible with this suggestion. With the assumption that excretion is coupled to the hydrolysis of 0.5 mol ATP per mol of propanediol, no net increase in the ATP generation would occur. Moreover, this could explain the rather low Y_{ATP} values calculated for steady-state conditions. As for the detrimental effect on the capacity to grow, this could not be attributed to toxic effects of the excretion products because similar concentrations of these products could be attained during steady-state growth conditions without any apparent effect on the cells.

181

The inevitable conclusion therefore seems to be that *K. aerogenes* grown under glycerol-limited conditions does not possess the metabolic flexibility and potential to cope with conditions of sudden glycerol excess. The putative presence of an active propanediol transport system needs to be established and the cause of the loss of viability under these conditions remains to be elucidated.

Conclusions

The natural habitats of many microbial species are characterized by significant nutritional fluctuations (Koch, 1971). It is therefore not surprising that, in the course of evolution, mechanisms have been developed to adapt rapidly to changes in the environment. Based on both ecophysiological (Tempest and Neijssel, 1978; Tempest *et al.*, 1985) and thermodynamic considerations (Westerhoff *et al.*, 1982, 1983), it has been proposed that metabolic flexibility can only be achieved by optimizing the catabolic flows for high rates and that such an optimization can only be effected at the cost of the efficiency of energy gain. Our results support this view. Although organisms are known to exist with an efficient but rigid metabolism [see, for example, Leegwater (1983) and Crabbendam *et al.* (1985)], we want to suggest that uncoupling of metabolism from energy generation is a general feature of microbial metabolism.

Uncoupling may occur either by a decreased efficiency of catabolism or by the presence of energy-spilling futile cycles (Neijssel *et al.*, 1978; Daldal and Fraenkel, 1983; Mulder *et al.*, 1986). Thus, energy dissipation may occur at various levels of metabolism and it will be clear that this phenomenon has a profound effect on the assessment and interpretation of yield values. It is a fact that calculations of the theoretical Y_{ATP}, based upon knowledge of cell composition and the energetics of the biosynthesis, invariably yield significantly higher values than those based on catabolic rates determined *in vivo* [see Stouthamer (1969, 1973, 1977, 1979); Hespell and Bryant (1979)]. Our results suggest that this may be due to the presence of uncoupling and energy spilling reactions [see also Tempest and Neijssel (1984)]. However, it is our belief that this anomaly can be understood in a broader sense when it is realized that growth cannot be described by a static model because it constitutes a complex of anabolic, catabolic and energy-dissipating *fluxes* [see also van Dam *et al.* (1986)].

The above-mentioned discrepancy is not only of scientific interest. For instance, it may explain the considerable heat output by actively metabolizing cells, an observation that should not be underestimated by technologists. It also illustrates that the use of microorganisms for industrial purposes can only be optimized when information is available about the actual *rates* of the metabolic processes within the cell. Therefore, it is essential that our knowledge about the relationships between the various intracellular flows and the environment are broadened.

Acknowledgements

We thank Carla Hoitink and Ed Buurman for their valuable contributions to the experiments presented here.

References

Bauchop,T. and Elsden,S.R. (1960) The growth of microorganisms in relation to their energy supply. *J. Gen. Microbiol.*, **23**, 157−169.

Beck,R.W. and Stugart,L.R. (1966) Molar growth yields in *Streptococcus faecalis* var. *liquefaciens*. *J. Bacteriol.*, **92**, 802−803.

Blackwood,A.C., Neish,A.C. and Ledingham,G.A. (1956) Dissimilation of glucose at controlled pH values by pigmented and non-pigmented strains of *Escherichia coli*. *J. Bacteriol.*, **72**, 497−499.

Brown,W.W. and Collins,E.B. (1977) End products and fermentation balances for lactic streptococci grown anaerobically on low concentrations of glucose. *Appl. Environ. Microbiol.*, **33**, 38−42.

Cooper,R.A. (1984) Metabolism of methylglyoxal in microorganisms. *Annu. Rev. Microbiol.*, **38**, 49−68.

Cooper,R.A. and Anderson,A. (1970) The formation and catabolism of methylglyoxal during glycolysis in *Escherichia coli*. *FEBS Lett.*, **11**, 273−276.

Crabbendam,P.M., Neijssel,O.M. and Tempest,D.W. (1985) Metabolic and energetic aspects of the growth of *Clostridium butyricum* on glucose in chemostat culture. *Arch. Microbiol.*, **142**, 375−382.

Daldal,F. and Fraenkel,D.G. (1983) Assessment of a futile cycle involving reconversion of fructose-6-phosphate during gluconeogenic growth of *Escherichia coli*. *J. Bacteriol.*, **153**, 390−394.

Driessen,M., Postma,P.W. and van Dam,K. (1987) Energetics of glucose uptake in *S. typhimurium*. *Arch. Microbiol.*, **146**, 358−362.

Forage,R.G. and Foster,M.A. (1982) Glycerol fermentation in *Klebsiella pneumoniae*: functions of the co-enzyme B_{12}-dependent glycerol and diol dehydratases. *J. Bacteriol.*, **149**, 413−419.

Gale,E.F. and Epps,H.M.R. (1942) The effect of the pH of the medium on the enzyme activities of bacterial (*Escherichia coli* and *Micrococcus lysodeikticus*) and the biological significance of the changes produced. *Biochem. J.*, **36**, 600−618.

Gottschalk,G. and Andreesen,J.R. (1979) Energy metabolism in anaerobes. In *International Review of Biochemistry. Microbial Biochemistry*. Quale,J.R. (ed.), University Park Press, Baltimore, Vol. 21, pp. 85−115.

Hansen,R.G. and Henning,U. (1966) Regulation of pyruvate dehydrogenase activity in *Escherichia coli* K12. *Biochim. Biophys. Acta*, **122**, 355−358.

Harder,W. and Dijkhuizen,L. (1983) Physiological reponses to nutrient limitation. *Annu. Rev. Microbiol.*, **37**, 1−23.

Herbert,D., Phipps,P.J. and Strange,R.E. (1971) Chemical analysis of microbial cells. In *Methods in Microbiology*. Norris,J.R. and Ribbons,D.W. (eds), Academic Press, London, Vol. 5b, pp. 209−344.

Hespell,R.B. and Bryant,M.P. (1979) Efficiency of rumen microbial growth; influence of some theoretical and experimental factors on Y_{ATP}. *J. Anim. Sci.*, **49**, 640.

Hommes,R.W.J., van Hell,B., Postma,P., Neijssel,O.M. and Tempest,D.W. (1985) The functional significance of glucose dehydrogenase in *Klebsiella aerogenes*. *Arch. Microbiol.*, **143**, 163−168.

Hopper,D.J. and Cooper,R.A. (1971) The regulation of *Escherichia coli* methylglyoxal synthase: a new control site in glycolysis. *FEBS Lett.*, **13**, 213−216.

Jones,C.W. (1977) Aerobic respiratory systems in bacteria. In *Microbial Energetics*. Haddock,B.A. and Hamilton,W.A. (eds), 27th Symposium of the Society of General Microbiology, Cambridge University Press, London, pp. 23−59.

Koch,A.L. (1971) The adaptive responses of *Escherichia coli* to a feast and famine existence. *Adv. Microb. Physiol.*, **6**, 147−217.

Kornberg,H.L. (1973) Fine control of sugar uptake by *Escherichia coli*. In *Rate Control of Biological Processes*. 27th Symposium on Environmental Biology, Cambridge University Press, Cambridge.

Leegwater,M.P.M. (1983) Microbial reactivity: its relevance to growth in natural and artificial environments. Thesis, University of Amsterdam.

Lin,E.C.C. (1976) Glycerol dissimilation and its regulation in bacteria. *Annu. Rev. Microbiol.*, **30**, 535−578.

McGee,D.E. and Richards,J.H. (1981) Purification and subunit characterization of propanediol dehydratase, a membrane-associated enzyme. *Biochemistry*, **20**, 4293−4298.

Meers,J.L., Tempest,D.W. and Brown,C.M. (1970) GOGAT oxidoreductase (NADP) in the synthesis of glutamate by some bacteria. *J. Gen. Microbiol.*, **64**, 187−194.

Muir,M., Williams,L. and Ferenci,T. (1985) Influence of transport energetization on the growth yield of *Escherichia coli*. *J. Bacteriol.*, **163**, 1237−1242.

Mulder,M.M., Teixeira de Mattos,M.J., Postma,P. and van Dam,K. (1986) Energetic consequences of multiple potassium uptake systems in *Escherichia coli*. *Biochim. Biophys. Acta*, **851**, 223−228.

Neijssel,O.M. and Tempest,D.W. (1975) The regulation of carbohydrate metabolism in *Klebsiella aerogenes* NCTC 418 organisms, growing in chemostat culture. *Arch. Microbiol.*, **106**, 251−258.

Neijssel,O.M. and Tempest,D.W. (1976) The role of energy spilling reactions in the growth of *Klebsiella aerogenes* NCTC 418 in aerobic chemostat culture. *Arch. Microbiol.*, **118**, 305−311.

Neijssel,O.M. and Tempest,D.W. (1979) The physiology of metabolite overproduction. In *Microbial Technology: Current State, Future Prospects*. Bull,A.T., Ellwood,D.C. and Radledge,C. (eds), 29th Symposium of the Society of General Microbiology, Cambridge University Press, London, pp. 53–82.

Neijssel,O.M., Sutherland-Miller,T.O. and Tempest,D.W. (1978) Pyruvate reductase and D-lactate dehydrogenase: a possible mechanism for avoiding energy conservation at site 1 of the respiratory chain in *Klebsiella aerogenes. Proc. Soc. Gen. Microbiol.*, **5**, 49.

Ramos,S. and Kaback,H.R. (1977) The relationship between the electrochemical proton gradient and active transport in *Escherichia coli* membrane vesicles. *Biochemistry*, **16**, 848–854.

Rosenberger,R.F. and Elsden,S.R. (1960) The yields of *Streptococcus faecalis* grown in continuous culture. *J. Gen. Microbiol.*, **22**, 726–739.

Stouthamer,A.H. (1969) Determination and significance of molar growth yields. In *Methods in Microbiology.* Norris,J.R. and Ribbons,D.W. (eds), Academic Press, London, Vol. 1, pp. 629–663.

Stouthamer,A.H. (1973) A theoretical study on the amount of ATP required for synthesis of microbial cell material. *Antonie van Leeuwenhoek*, **39**, 545–565.

Stouthamer,A.H. (1977) Energetic aspects of the growth of microorganisms. In *Microbial Energetics.* Haddock, B.A. and Hamilton,W.A. (eds), 27th Symposium of the Society of General Microbiology, Cambridge University Press, London, pp. 285–315.

Stouthamer,A.H. (1979) The search for correlation between theoretical and experimental growth yield. In *International Review of Biochemistry. Microbial Biochemistry.* Quayle,J.R. (ed.), University Park Press, Baltimore, Vol. 21, pp. 28–47.

Stouthamer,A.H. and Bettenhausen,C.W. (1975) Determination of the efficiency of oxidative phosphorylation in continuous cultures of *Aerobacter aerogenes. Arch. Microbiol.*, **102**, 187–192.

Streekstra,H., Teixeira de Mattos,M.J., Neijssel,O.M. and Tempest,D.W. (1987a) Overflow metabolism during anaerobic growth of *klebsiella aerogenes* NCTC 418 on glycerol and dihydroxyacetone in chemostat culture. *Arch. Microbiol.*, in press.

Streekstra,H., Teixeira de Mattos,M.J., Buurman,E.T., Hoitink,C.W.G., Neijssel,O.M. and Tempest,D.W. (1987b) Fermentation shifts and metabolic reactivity during anaerobic carbon-limited growth of *Klebsiella aerogenes* NCTC 418 on fructose, gluconate, mannitol and pyruvate. *Arch. Microbiol.*, submitted.

Teixeira de Mattos,M.J. and Tempest,D.W. (1983) Metabolic and energetic aspects of growth of *Klebsiella aerogenes* NCTC 418 on glucose in anaerobic chemostat culture. *Arch. Microbiol.*, **134**, 80–85.

Teixeira de Mattos,M.J., Streekstra,H. and Tempest,D.W. (1984) Metabolic uncoupling of substrate level phosphorylation in anaerobic glucose-limited chemostat cultures of *Klebsiella aerogenes* NCTC 418. *Arch. Microbiol.*, **139**, 260–264.

Tempest,D.W. and Neijssel,O.M. (1978) Ecophysiological aspects of microbial growth in aerobic nutrient-limited environments. In *Advances of Microbial Ecology.* Alexander,M. (ed.), Plenum Press, London, Vol. 2, pp. 105–153.

Tempest,D.W. and Neijssel,O.M. (1984) The status of Y_{ATP} and maintenance energy as biologically interpretable phenomena. *Annu. Rev. Microbiol.*, **38**, 459–486.

Tempest,D.W., Neijssel,O.M. and Zevenboom,W. (1983) Properties and performance of microorganisms in laboratory culture; their relevance to growth in natural ecosystems. In *Microbes in their Natural Environment.* Slater,J.H., Whittenbury,R. and Wimpenny,J.W.T. (eds), 34th Symposium of the Society of General Microbiology, Cambridge University Press, London, pp. 119–152.

Tempest,D.W., Neijssel,O.M. and Teixeira de Mattos,M.J. (1985) Regulation of carbon metabolism growing in chemostat culture. In *Environmental Regulation of Microbial Metabolism.* Kulaev,J.S., Dawes,E.A. and Tempest,D.W. (eds), 23rd FEMS Symposium, Academic Press, London, pp. 53–69.

Thauer,R.K., Jungermann,K. and Decker,K. (1977) Energy conservation in chemotrophic anaerobic bacteria. *Bacteriol. Rev.*, **41**, 100–180.

Thomas,T.D., Ellwood,D.C. and Longyear,V.M.C. (1979) Change from homo- to heterolactic fermentation by *Streptococcus lactis* resulting from glucose limitation in anaerobic chemostat cultures. *J. Bacteriol.*, **138**, 109–117.

Tran Din,K. and Gottschalk,G. (1985) Formation of D(−)-1,2-propanediol and D(−)-lactate from glucose by *Clostridium sphenoides* under phosphate limitation. *Arch. Microbiol.*, **142**, 87–92.

van Dam,K., Mulder,M.M., Teixeira de Mattos,M.J. and Westerhoff,H. (1986) A thermodynamic view of bacterial growth. In *Physiological Models. CRC Series Mathematical Models of Microbiology*, in press.

Wallace,R.J. (1978) Control of lactate production by *Selenomonas ruminantium*; homotropic activation of lactate dehydrogenase by pyruvate. *J. Gen. Microbiol.*, **107**, 45–92.

Westerhoff,H., Lolkema,J.S., Otto,R. and Hellingwerf,K.J. (1982) Thermodynamics of bacterial growth. The phenomenological and the mosaic approach. *Biochim. Biophys. Acta*, **683**, 181–220.

Westerhoff,H., Hellingwerf,K.J. and van Dam,K. (1983) Thermodynamic efficiency of microbial growth is low, but optimal for maximal growth rate. *Proc. Natl. Acad. Sci. USA*, **80**, 305–309.

CHAPTER 11

Industrial gases containing carbon monoxide as substrates for biotechnology

E.WILLIAMS, J.COLBY[1], G.W.LOGAN[1] and C.M.LYONS

Microbial Technology Group, Department of Microbiology, The Medical School, Framlington Place, Newcastle upon Tyne NE2 4HH and [1]North East Biotechnology Centre, Department of Biology, The Polytechnic, Sunderland SR1 3SD, UK

Introduction

Approximately half of the energy content of the recoverable reserves of all fossil fuels in the world are in the form of coal, lignite and peat, 12% are in the form of crude oil, 6% as natural gas and the remainder as oil shales and bituminous sands (Gibbs and Greenhalgh, 1983). Although the upsurge in oil prices stimulated the discovery of new oil and gas reserves, about 60% of the oil reserves are found in a few Middle East countries. At present rates of consumption the world's oil and gas reserves are expected to be depleted in a few decades (Kölling and Schnur, 1982). In contrast, coal is more evenly distributed, with large reserves in North America, China, the Soviet Union, Australia, Western Canada and the European Community. It has been estimated these reserves will meet current demands for 250 years (Gibbs and Greenhalgh, 1983).

The conversion of plant material to coal passes through several phases. Peats and soft lignite coals are firstly formed followed by hard lignite coals and hard coals where the degree of carbonization increases. All coals consist of carbon, hydrogen, oxygen, nitrogen, and sulphur, their proportions varying with age and degree of carbonization (*Table 1*). The organic content of coals consist of mostly high molecular weight materials such as bitumens (waxes and resins) and humic acids. Hard coals consist mainly of aromatic compounds, whereas lignites contain aliphatic compounds as well (Kölling and Schnur, 1982).

The organic chemical industry was originally based on coal products such as tar dyes (azo, alizarine, indanthrine), pharmaceuticals and pesticides. The plastics polystyrene and polyurethanes are based on feedstocks that were once obtained only from coal. Ammonia is produced as a by-product of coking of coal and was for many years an important process for the production of artificial fertilizers. This process was replaced by synthetic processes using hydrogen and nitrogen; the hydrogen was originally obtained from coal, but is now mostly obtained from natural gas or oil. Calcium carbide and acetylene were other important products of coal chemistry. However, the advantages of liquid (oils) and gaseous (natural gas) feedstocks, and the increased demand for chemical feedstocks such as benzene and phenols, encouraged a switch to these

Table 1. Composition of peat and coals.

	% Water content	% Carbon	% Hydrogen	% Oxygen	% Nitrogen	% Sulphur
Peat	80−90	50−60	5−8	30−45	1−4	0.1−1
Soft lignite coal	55−63	65−70	5−8	20−30	0.5−1.5	0.5−3
Hard lignite coal	30−40	70−73	5−8	16−25	0.5−1.5	0.5−3
Anthracite	<1	>91.5	<3.75	<2.5	1−1.7	0.6−1.7

Data from Kölling and Schnur (1982)

Table 2. Typical gas compositions (vol. %) of various gasification processes.

Process	CO	H_2	CO_2	N_2
Leuna Slagging	64.6	28.7	5.7	1.0
Lurgi Pressure	15.7−23	25.1−39.1	14−32.4	0.7−40
Lurgi Slagging	58.0	25.6	6.3	1.0
Winkler	30−50	35−46	13−25	0.5−1.5
Koppers−Totzek	57.5	30.7	10.5	1.2
Rummel−Otto	56.8	28.0	14.0	1.0
Ruhrgas Fluidized Bed	22.8	8.0	5.1	64.1
Babcock and Wilcox/Dupont	40.2	39.4	16.2	3.0
Pintsch−Hillebrand	28.0	56.0	14.0	1.0
Koppers Recycle	28.1	56.1	13.0	1.2
Lurgi−Ruhrgas	32.2	53.4	10.0	0.5
Coalcon	<39.0	48−66	3−28	0
Cogas	31.2	57.9	6.6	0.3
Hygas	7.4−21.3	22.5−24.2	0.5−1.0	<1.0
Bi-Gas	29.4	32.1	21.5	0.6

Data from Baron et al., 1982.

materials in the middle of the twentieth century. Oil and natural gas were initially used to produce synthesis gas, a mixture of carbon monoxide (CO) and hydrogen (H_2), and later for the manufacture of olefins and aromatic chemicals. The use of coal was then mainly restricted to supplying aromatic compounds as by-products of coke production (Kölling and Schnur, 1982).

Synthesis gas or syn gas describes mixtures of CO and H_2 in various molar ratios from 2:1 to 1:3, depending on the production process used. It may be produced from any material containing carbon and hydrogen (Graboski, 1984), but the most important industrial sources are:

(i)   from steam and methane at 850°C and 30 bar pressure with a nickel catalyst:
$$CH_4 + H_2O = CO + 3H_2$$
(ii)   by the partial oxidation of fuel oil:
$$C_nH_{2n} + {}^n/_2O_2 = {}_nCO + {}_nH_2$$
(iii)   by coal gasification:
$$C + H_2O = CO + H_2$$
$$C + {}^1/_2O_2 = CO$$

Although synthesis gas is currently produced predominantly from natural gas and oil, the economic advantages of these feedstocks over coal may change as oil and gas reserves become depleted. Improvements in coal gasification techniques and technical improvements in underground gasification is expected to lead to the availability of large amounts of synthesis gas at low costs (Sheldon, 1983).

Gasification of coal is more difficult than natural gas or oil because of its solid form, high C/N ratio and high ash content (Graboski, 1984). It is accomplished by reacting hard or lignite coal with a gasification agent such as air, oxygen, steam, carbon dioxide or hydrogen at temperatures greater than 750°C. The resulting gas mixtures contain CO and H_2 as the principal components together with methane, nitrogen and hydrogen sulphide; the ratios of the constituent gases varying with the gasification process used. Typical gas compositions of various processes are presented in *Table 2*. The heat for gasification may either be supplied by combusting part of the feedstock with gasification agents containing oxygen, or may be supplied externally, using residual coke or heat from nuclear power (Baron *et al.*, 1982; van Heek and Juntgen, 1982). There are wide variations in the particle size of coal, resulting in fluidized bed (<3 mm particle size) or fixed bed (>3 mm particle size) operation. The main first-generation commercial processes are the Lurgi, Koppers–Totzek and Winkler methods. These have largely been replaced by the Texaco and Koppers–Babcock–Wilcox processes. Recently the British Gas Corporation have developed a slagging Lurgi gasifier which operates as a fixed bed with oxygen lances at the bottom (Graboski, 1984).

The widespread availability of synthesis gas from coal may depend on improvements in underground gasification of coal deposits. This may lead to abundant and secure supplies of synthesis gas by exploiting offshore seams or those at depths of $1000-2000$ m. The technique consists of forcing the gasifying agent through the coal seam under pressure around a circuit developed through boreholes (Gibbs and Greenhalgh, 1983). In recent trials in Belgium the size of the natural fissures and the reaction rate was increased by passing compressed air through the seam. Although most processes use air or oxygen as gasifier, Germany has a programme for producing synthesis gas (for piping into the current gas network) from lignite deposits using hydrogen as gasifier.

The production of chemicals from synthesis gas follows several routes (Frohning *et al.*, 1982; Sheldon, 1983; Knifton *et al.*, 1984). It is used on a large scale for the production of methanol and ammonia:

$$CO + 2H_2 = CH_3OH$$
$$N_2 + 3H_2 = 2NH_3$$

Methanol is used as an industrial solvent and as raw material for a variety of industrial chemicals such as formaldehyde, methylamines, acetic acid, methyl chloride, dimethyl terephthalate and methyl crylate, and is used as the carbon and energy source for single cell protein production ('Pruteen', Imperial Chemicals Industry Ltd). Hydrocarbons may be synthesized by the iron-catalysed Fischer–Tropsch process. This process has been operated for many years by the Sasol Company in South Africa for the production of gasoline. There is also wide interest in using the Fischer–Tropsch process for producing ethylene:

$$2CO + 4H_2 = CH_2{=}CH_2 + 2H_2O$$

Other products of interest are ethylene glycol, vinyl acetate and methane (Seglin *et al.*, 1975):

$$2CO + 3H_2 = HOCH_2CH_2OH$$
$$2CO + 2H_2 = CH_3COOH$$
$$4CO + 5H_2 = CH_2{=}CHOA_c + H_2O$$
$$CO + 3H_2 = CH_4 + H_2O$$

Although these processes are expected to increase in importance, there are potentially biological routes for the conversion of synthesis gas to useful products, especially high value pharmaceuticals.

The properties and occurrence of carbon monoxide

The gas mixtures in *Table 2* provide two main energy sources for microbial growth, H_2 and CO. The aerobic hydrogen bacteria are well known, but they will not grow on gases containing high concentrations of carbon monoxide (Kesler *et al.*, 1978; Volova *et al.*, 1980; Cypionka and Meyer, 1982, 1983).

Carbon monoxide is perhaps an unusual substrate to be considered for bacterial growth. It is a colourless, odourless, tasteless, explosive and extremely toxic gas. It has an ignition temperature in air of 700°C within the flammable limits of 12−75%. Solubility in water is low (3.3 ml/100 ml water at 0°C and 2.3 ml/100 ml water at 20°C) and this limits its availability as a substrate for fermentation processes, particularly at high temperatures (Sheldon, 1983; Hughes, 1985).

Carbon monoxide is produced from many biological and anthropogenic sources (Kim and Hegeman, 1983). Anthropogenic production of CO is varied, but mainly results from incomplete combustion processes. Blast furnace gas (25−30% CO), automobile exhaust (0.5−12% CO) and faulty domestic heating systems are the main sources. Man also produces small amounts from the oxidation of haemoglobin by microsomal haem oxygenase; this is released through the lungs and may accumulate in enclosed environments. Man is also exposed to the poisonous effects of CO by smoking cigarettes (2−5% CO), cigars and pipe tobacco (5−14% CO) (Williams and Colby, 1986).

Carboxydotrophic bacteria

A large number of bacterial species belonging to several genera have been shown to utilize CO as an energy source. The aerobic bacteria are collectively known as the carboxydotrophic bacteria. Four main groups have been described (Williams *et al.*, 1986). The mesophilic Gram-negative bacteria are rods with curved or rounded ends, have optimum growth temperatures of 30°C and do not produce spores. The well-characterized species are, *Pseudomonas carboxydovorans*, *P. carboxydohydrogena*, *P. carboxydoflava*, *P. compransoris*, *P. gazotropha* and *Alcaligenes carboxydus*. All except *Alcaligenes carboxydus* are capable of growing with H_2 as well as CO as energy source. They are all facultative autotrophs, capable of utilizing a limited range of multicarbon compounds.

The Gram-positive mesophilic bacteria are represented by *Arthrobacter* strain 11/x. This was isolated as a hydrogen bacterium, but was found to grow well with CO (Meyer and Schlegel, 1983). It has a variable rod/coccus morphology, does not produce spores and has an optimum growth temperature of 37°C. *Bacillus schlegelii* a hydrogen bacterium was found to grow with CO as sole carbon and energy source (Meyer and Schlegel, 1983). Four more strains of carboxydotrophic, thermophilic bacilli were subsequently isolated. All strains produced Gram-variable, long rod-shaped cells with terminal, spherical endospores which distended the sporangium. The optimum growth temperatures are 65 – 70°C.

In our own laboratories we have developed several selective isolation techniques for novel types of carboxydotrophic bacteria. We have recently described a Gram-positive thermophilic mycelial bacterium, *Streptomyces* G26, containing LL-diaminopimelic acid and glycine in the cell wall. It is moderately thermophilic, with an optimum growth temperature of 50°C (Bell *et al.*, 1985). We have also reported the isolation of a novel Gram-negative, thermophilic carboxydotroph, *Pseudomonas thermocarboxydovorans* (Lyons *et al.*, 1984).

Growth physiology of *Pseudomonas thermocarboxydovorans*

Autotrophic and heterotrophic growth

P. thermocarboxydovorans (strain C2, NCIMB11893) was isolated from sewage (a primary, aerobic settlement tank) by liquid enrichment in a mineral medium at 45°C under an atmosphere containing 25%(v/v) CO, 25%(v/v) H_2 and 50%(v/v) air (Lyons *et al.*, 1984). This moderately thermophilic carboxydotroph is capable of autotrophic growth on CO alone, or on a mixture of CO and H_2, although in the latter case the H_2 is not utilized. *P. thermocarboxydovorans* is incapable of growth as a hydrogen bacterium, apparently because it cannot oxidize H_2 (Lyons *et al.*, 1984). The organism also grows as a heterotroph and can utilize, as carbon and energy source, a rather limited range of organic and amino acids. One-carbon and two-carbon compounds are not utilized, nor are sugars. Ammonia, nitrate, hippurate, urea and some amino acids serve as sources of nitrogen, but dinitrogen gas and methylamine are not used.

Growth rates and yields

P. thermocarboxydovorans was the first aerobic carboxydotroph isolated capable of rapid growth with CO as sole carbon and energy source (Lyons *et al.*, 1984). Doubling times of about 3 h are obtained in defined mineral medium under an atmosphere of 50%(v/v) CO in air. Subsequently, similar growth rates on CO have been observed with several *Bacillus* strains, including *B. schlegelii* (Kruger and Meyer, 1984), and with autotrophic actinomycetes including *Streptomyces* G26 (Bell *et al.*, 1985). Heterotrophic growth of *P. thermocarboxydovorans* on 0.2% sodium pyruvate is more rapid with doubling times of about 1 h, dropping to 24 min on augmenting the medium with low concentrations (0.05 – 0.1%) of organic supplements such as yeast extract, casamino acids or nutrient broth. Higher concentrations of these complex sources are inhibitory. Growth on CO, oxidation of CO, and CO_2 production are all slower in the presence of concentrations of oxygen higher than 15% (v/v). Growth is most rapid at temperatures between 45°C and 55°C with a lower limit at 42°C, where the doubling time is 5 h, and a higher limit at 65°C.

Growth stoichiometry during batch growth on CO was determined using the constant pressure apparatus shown in *Figure 1*. The results obtained are shown in *Figure*

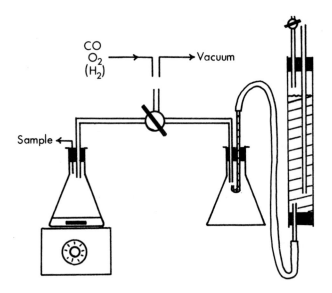

Figure 1. Constant pressure apparatus for measuring growth stoicheiometry during batch growth on CO. The gaseous volume of the apparatus was initially 1205 ml and the apparatus was gassed with a mixture of 33% CO in air. The experiment was set up in a 45°C incubator and the apparatus was surrounded with insulating material to minimize temperature fluctuations. Following inoculation, samples of gas and culture were removed at intervals for gas analysis by thermal conductivity gas chromatography and cell density measurement; volume changes were also noted. An uninoculated control apparatus was treated similarly and the experimental values adjusted accordingly.

GROWTH STOICH. DETN.

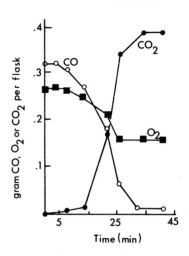

Figure 2. Growth stoicheiometry of *Pseudomonas thermocarboxydovorans* on CO.

2 and are consistent with the following growth equation:

$$CO + 0.357 \text{ g } O_2 = 0.097 \text{ g cells} + 1.304 \text{ g } CO_2$$

Comparable data for the mesophile *P. carboxydovorans* growing on CO at 30°C (quoted

Table 3. Enzyme activities in *Pseudomonas thermocarboxydovorans* grown in chemostat culture.

	Growth substrate				
	CO			Pyruvate	
Nutrient limitation *Enzymes*	NH_3	O_2	C	NH_3	C
CO oxidoreductase	944	1293	2349	0	720
Ribulose bisphosphate carboxylase	61	68	34	4	8
Hydroxypyruvate reductase	55	81	64	51	nd
Phosphoglycollate phosphatase	43	190	120	52	47

Enzyme activities are in nmol substrate transformed or product formed per min per mg protein. The soluble fraction of ultrasonic extracts was used for all assays except for the dye-linked hydrogenase assay where both soluble and particulate fractions were tried.

by Meyer, 1980) indicate the following stoichiometry.

$$CO + 0.457 \text{ g } O_2 = 0.164 \text{ g cells} + 0.836 \text{ g } CO_2$$

This indicates a more efficient conversion of CO into cell carbon in *P. carboxydovorans*, although it should be borne in mind that the growth rate of the latter organism is considerably lower ($t_D = 20$ h compared with 3 h).

Regulation of autotrophic metabolism

Table 3 shows the specific activities of certain key enzymes of CO oxidation and assimilation pathways in soluble extracts prepared from *P. thermocarboxydovorans* grown in chemostat culture on CO or on sodium pyruvate.

CO oxidase is the enzyme responsible for the oxidation of CO and hence for the supply of electrons to the respiratory electron transport chain for ATP synthesis, the supply of CO_2 for cell carbon, and of NADPH to drive the Calvin cycle. In *P. thermocarboxydovorans* it is clearly an inducible enzyme with highest activities observed in CO-limited culture. A reasonably high specific activity of the enzyme was also observed in carbon-limited culture with pyruvate as carbon source, suggesting that regulation of enzyme synthesis involves repression by organic compounds present in the growth medium. Ribulose bisphosphate carboxylase is present in high specific activity in autotrophic cultures but not when pyruvate is the carbon source. Hydroxypyruvate reductase has been implicated in the assimilation of phosphoglycollate arising in autotrophs as a consequence of the ribulose bisphosphate oxygenase activity of ribulose bisphosphate carboxylase (Bamforth and Quayle, 1977; Taylor, 1977; Beudeker *et al.*, 1981). The presence of relatively high specific activities of both this enzyme (1) and of phosphoglycollate phosphatase (2) in *P. thermocarboxydovorans* suggests that the glycine−serine pathway shown below may operate in this organism.

Phosphoglycollate $\xrightarrow{(2)}$ Glycollate -- Glyoxylate -- Glycine

3 - Phosphoglycerate -- Glycerate $\xrightarrow{(1)}$ Hydroxypyruvate -- Serine

Table 4. Poly-β-hydroxybutyrate production by *Pseudomonas thermocarboxydovorans*.

Carbon source	Growth mode	Limiting nutrient	PHB content
Pyruvate	Batch	–	0.4
	Continuous	oxygen	9.1
	Continuous	ammonia	31.9
CO	Batch	–	0.0
	Continuous	oxygen	0.9
	Continuous	ammonia	20.9

Chemostat cultures were operated at a dilution rate of 0.05/h. The PHB contents of freeze-dried organisms grown under various conditions are expressed as mg PHB per 100 mg dry weight.

Poly-β-hydroxybutyrate (PHB) biosynthesis

P. thermocarboxydovorans forms PHB as a storage material as can be readily demonstrated by staining older cultures with Nile blue A. The accumulation of this polymer has been investigated under batch and chemostat culture conditions with either CO or sodium pyruvate as carbon and energy source. The results are shown in *Table 4*. Significant accumulation [21−32% (w/w)] occurs only in chemostat culture under ammonia limitation with either carbon source. Smaller amounts are observed under oxygen limitation. PHB is being produced commercially by Marlborough Biopolymers Ltd (an ICI subsidiary) as a biopolymer alternative (BIOPOL) to the more traditional plastics such as polypropylene and polyvinylchloride. Although presently not competetive in price, BIOPOL has significant advantages, such as its biodegradability and biocompatability, and may become commercially important in medicine and in providing biodegradable packaging materials for our increasingly environmentally conscious society. The current commercial process involves an aerobic fermentation using a carbohydrate substrate ('plastic from sugar'). However, the source organism for the commerical process, *Alcaligenes eutrophus*, is also capable of growth on hydrogen and carbon dioxide but does not grow on industrial gases containing carbon monoxide. A combination of this organism (or rather a thermophilic equivalent) with a PHB-producing carboxydotroph such as *P. thermocarboxydovorans* might be ideal for the bioconversion of synthesis gas into bioplastic ('plastic from coal' by a biological route).

Optimization of a CO batch fermentation

In an aerobic fermentation, the mass transfer of molecular oxygen from the gas to the liquid phase is crucial. In order to maximize the transfer of oxygen during the growth of *P. thermocarboxydovorans*, a series of optimization experiments were carried out on a laboratory-scale fermenter (Bioengineering KLF 2000) containing 2 litres of mineral medium; no anti-foaming agent was used. The mass transfer rate coefficient was measured using an unsteady state oxygen electrode technique. Two types of design analysis were compared.

The single factor method is the traditional approach used by biologists. All but one of the experimental parameters are held constant whilst the single remaining factor is varied over the experimental response area of interest until the optimum is found. A new factor is then chosen and the remaining parameters held constant and the experiment repeated. The optima for all the factors are 'pooled' and the resultant response is regarded as the optimum for all parameters.

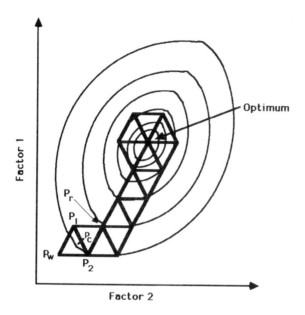

Figure 3. Hypothetical response surface.

This technique has a number of disadvantages including the cost of the large number of experiments required and the difficulty of selecting the intervals at which each factor should be tested. It is only too easy to miss the optimum set of conditions completely. These drawbacks can be overcome if two or more factors can be varied simultaneously as is possible using the simplex method.

This work was done in collaboration with R. Nicol and Dr P. Smith of the Department of Mathematics and Computer Studies at Sunderland Polytechnic. In our study, the simplex method of Nelder and Mead (1965) was used. This is a sequential method and should not be confused with the simplex method of linear programming. The simplex can be defined as a figure with one more vertex than the number of dimensions (= experimental factors or parameters). Thus for two parameters the simplex is a triangle, for three parameters it is a tetrahedron. It is easiest to envisage a two-parameter system to illustrate the principle (*Figure 3*). Each vertex of the simplex corresponds to a set of experimental conditions the response (in this case the *K*la) of which is then evaluated. The point with the worst response (Pw) is then selected and reflected through the centroid (Pc) of the remaining points. The response at the new point (Pr) is then evaluated experimentally. If this response (*K*la) is higher, then Pw is abandoned and Pr incorporated into the new simplex. In this way the simplex 'flips' itself over towards the optimum. For more than three dimensions, an algebraic resolution of 'flipping' the vertex points is used. The simplex can be expanded and then contracted as the optimum is approached in order to improve the accuracy of the method and/or minimise the number of experiments required.

In optimizing the CO fermentation the starting point was the fermentation conditions used prior to this study which gave the *K*la for oxygen of 16.9 /h. *Table 5* gives data for these original fermentation conditions together with the final values obtained by using the single factor and simplex methods. The single factor optimum was constrain-

Table 5. Optimization of fermentation conditions with carbon monoxide.

System	r.p.m.	pH	v.v.m	Tempera- ture (°C)	Kla (h)	Fermentation time (h)	Dry weight g/l
Original	900	7.0	0.1	50	16.9	68	0.30
Single factor	1700	7.0	0.2	50	71.7	34	0.28
Simplex	1414	7.0	0.2	46	94.0	28	0.30

The original fermentation conditions represent the starting point for the optimization exercise. The 'optimum' conditions were determined by single-factor or simplex analysis. Fermentations were then run using these three sets of conditions with the results shown.

ed by the mechanical tolerance levels of the equipment used; one advantage of the simplex is that these boundaries can be defined at the outset. In using the simplex method, four parameters were considered: pH, temperature, agitation speed (r.p.m.) and aeration rate (v.v.m.). The results obtained for the simplex method are shown after 15 'flips' of the system. The Kla steadily increased until a maximum value of 94.0/h was obtained. The simplex was then terminated since the optimum response area was resolved.

The results of the optimization exercise were tested by running fermentations under three sets of conditions, representing:

(i)   the original fermentation parameters;
(ii)   the optimum parameter set determined by single-factor analysis within the constraints of mechanical performance; and
(iii)   the optimum parameter set determined by the simplex method.

The results are shown in *Table 5*. In each case, the cells were harvested when the dry weight concentration of the culture was 0.3 g/l.

In comparison with single factor analysis, the simplex method yielded optimum conditions accounting for a reduction in fermentation time coupled to a reduction in the total power requirement for the fermentation. This was achieved with only 30 individual experiments whereas the single factor method required 54 providing a substantial reduction in time and resources.

We are currently extending this work by scaling-up to a 40-l CO fermentation and by considering other factors such as the mass transfer of CO.

Carbon monoxide oxidation

P. thermocarboxydovorans contains a CO oxidase that is a molybdenum hydroxylase of molecular weight 233 000 ± 15 000 containing iron-sulphur and FAD prosthetic groups (Turner *et al.*, 1984). The enzyme has now been purified to homogeneity in our laboratories and 1 mol of the enzyme has been shown to contain $1.6-2.1$ FAD, 6.9 Fe, 6.9 acid-labile sulphide and $0.5-0.7$ Mo. The absorption spectrum of the enzyme has a broad double peak at about 420 nm, a shoulder at 545 nm and a trough at 400 nm. The enzyme catalyses the oxidation of CO to CO_2 with the reduction of an added electron acceptor with a E'_0 in the range $+0.011V$ to $+0.429V$. Nicotinamide nucleotides, ferredoxins, flavins and viologens are not used. Notable features of the enzyme are:

(i)   its high degree of specificity for CO;
(ii)   its high sensitivity to CO (K_m 1 μM); and
(iii)   its relatively high thermal stability (Colby *et al.*, 1985).

There has been some discussion in the literature about the subunit structure of CO oxidases. Our work indicates that the enzyme from *P. thermocarboxydovorans* is a hexamer of three non-identical subunits (i.e. an $L_2M_2S_2$ structure) of molecular weight $87\,000 \pm 2060$, $29\,000 \pm 1500$ and $21\,000 \pm 3000$. We find no evidence for the L_3 configuration proposed by Meyer and Rohde (1984).

Soluble extracts of *P. thermocarboxydovorans*, prepared by centrifuging crude ultrasonic or French pressure-cell homogenates at $100\,000$ g for 1 h, contain all the CO oxidase activity. Nevertheless, the enzyme is thought to be associated with the inner face of the cell membrane *in vivo*. Meyer and Rohde (1984), using *P. carboxydovorans*, demonstrated that CO oxidase is loosely attached to the cytoplasmic membrane being released into the cytoplasm on the onset of stationary phase. CO oxidase is thought to interact with the electron transport chain at the level of ubiquinone or cytochrome *b*. Carboxydotrophic bacteria have been shown to contain normal electron transport conponents such as *a*-, *b*- and *c*-type cytochromes (Cypionka and Meyer, 1983) despite their ability to grow in the presence of high nutritional levels of CO. *P. carboxydovorans* contains UQ_{10} as its only quinone (Meyer and Schlegel, 1980). The electron transport chain of *P. carboxydovorans* is branched with heterotrophic (CO-sensitive) and autotrophic (CO-insensitive) branches diverging at the level of UQ_{10} or a *b*-type cytochrome (Cypionka and Meyer, 1983). The terminal oxidase of the autotrophic branch is thought to be cytochrome *o* ($=b_{563}$).

Genetics of *Pseudomonas thermocarboxydovorans*

Mutagenesis

Mutation induction in obligate methanotrophic bacteria is unusual. Repair-dependent mutagens, such as u.v. radiation, ionizing radiations and methyl methanesulphonate (MMS) are extremely poor mutagens, and repair-independent mutagens, such as N'-methyl-N'-nitro-N-nitrosoguanidine (MNNG) and ethyl methanesulphonate induce mutations at low levels (Harwood *et al.*, 1972; Williams and Bainbridge, 1976; Williams *et al.*, 1977; Williams and Shimmin, 1978). *P. thermocarboxydovorans* shows a normal response to mutagens of both the repair-dependent and repair-independent types, and a range of mutants including auxotrophic and autotrophic-negative mutants have been isolated.

Several forward markers have been used to estimate the frequency of induced mutation after various treatments with mutagens. Streptomycin was the prefered marker as it was stable at $50\,^\circ$C, the spontaneous frequency was low but measurable and the resistant mutants had a minimum inhibitory concentration 10-fold higher than the wild-type and were stable.

The loss of viability in response to u.v. radiation was exponential (*Figure 4*). The survival curve was shoulderless with a break, giving an overall sigmoidal appearance to the curve. This break was mirrored in the mutation induction curve. The position and extent of the break varied with the medium used; on pyruvate minimal medium the break was quite marked and occurred at about 3% survival, on enriched medium the break was less prominent and occurred at lower survival levels. The slope of the survival curves for cells plated on enriched medium were also much steeper than on pyruvate minimal medium alone. This increase in apparent sensitivity appeared to be related to the growth rate. The mutation frequency also increased on enriched medium

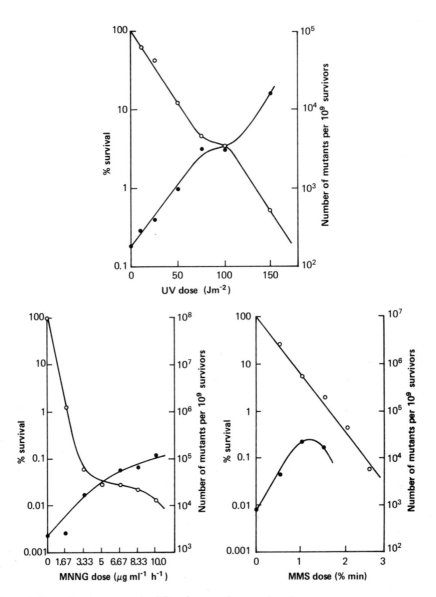

Figure 4. Survival and mutagenesis of *Pseudomonas thermocarboxydovorans*.

(*Table 6*). There was no level of survival or mutagen dose which resulted in the production of a peak of mutants, that is the measured mutation frequency was inversely related to the percentage survival (*Figure 4*). *P. thermocarboxydovorans* has an efficient photoreactivation system which promotes survival in the presence of visible or blue light (*Table 7*).

Exposure of *P. thermocarboxydovorans* to MMS resulted in an exponential loss of viability with dose rate (*Figure 4*). The slope of the curve was approximately the same whether the cells were plated on pyruvate minimal medium with or without yeast ex-

Table 6. Ultraviolet radiation survival and mutation in *Pseudomonas thermocarboxydovorans.*

U.v. dose J/m^2	Pyruvate minimal medium		+ 0.1% Yeast extract	
	% Survival	Mutation frequency[a]	% Survival	Mutation frequency[a]
10	100	1.79×10^2	100	1.5×10^2
25	61.9	2.81×10^2	6.34	4.5×10^4
50	42.6	4.05×10^2	0.88	2.36×10^5
100	3.42	3.01×10^3	4.4×10^{-2}	6.0×10^6
150	0.52	1.34×10^4	4.1×10^{-3}	6.0×10^7

[a]Mutation frequency is expressed as the number of mutants per 10^9 survivors.

Table 7. Photoreactivation of ultraviolet light in *Pseudomonas thermocarboxydovorans.*

U.v. dose J/m^2	No photoreactivation % Survival	Photoreactivation % Survival
0	100	100
10	61.9	100
25	42.6	100
50	12.4	82.4
100	3.5	75.0
150	0.52	64.7

tract [0.1% (w/v)] enrichment. Unlike u.v. radiation, the mutation induction curve peaked. The maximum increase in mutation frequency was approximately 30 times the spontaneous level, and corresponded to a dose of approximately 1% min.

Treatment of isolate *P. thermocarboxydovorans* with MNNG produced a shoulderless, exponential survival curve with a resistant tail portion at high doses (*Figure 4*). The mutation induction curve plateaued at an induced mutation frequency 50 times the spontaneous frequency, corresponding to a dose of 10 μg/ml/h.

Despite the very large increase in mutation frequency, in particular when the mutated cells were grown on medium enriched with 0.1% (w/v) yeast extract, no auxotrophic or autotrophic mutants were isolated from more than 1000 colonies tested at each of two u.v. doses; 100 J/m^2 and 250 J/m^2. The apparent increase in mutation frequency, above the spontaneous level, for the latter dose was some 400 000-fold. Treatment of *P. thermocarboxydovorans* with MMS at a dose of 1% min resulted in the isolation of auxotrophic mutants at a frequency of 0.4%. Using MNNG auxotrophic and autotrophic-defective mutants were isolated at frequencies of 1% and 5%, respectively, from 3000 colonies.

The transfer of the transposon Tn5 resulted in the isolation of auxotrophic mutants at a frequency of approximately 0.5%. Reversion frequencies of these mutants were not calculated but from empirical observations they were less stable than mutants derived from chemical mutagenesis.

Biotechnological exploitation of microbial CO oxidation

There are four areas where carboxydotrophic bacteria may be exploited for biotechnology.

Table 8. Menaquinone composition of Gram-positive filamentous carboxydotrophs.

Strain	Menaquinone isoprenologue MK9			
	H_2	H_4	H_6	H_8
1	−	+ +	+ + +	+ +
2	−	+	+ + +	+ +
3	−	+ +	+ + +	+ +

H_2, H_4, H_6, H_8 denotes the degree of hydrogenation of the isoprenologue. The main component is indicated by + + +, any component greater than 50% of the main peak by + + and other significant components by +.

Table 9. Fatty acid composition of Gram-positive filamentous carboxydotrophs.

Strain no.	i14	14:0	i15	ai15	15:0	a	b	i16	16:1	16:0	c	d	i17	ai17	17:0	i18	18:1	18:0
1	−	1.7	−	3.1	−	−	19.0	−	17.0	−	−	2.2	16.7	13.1	12.5	17.6	8.0	
2	−	1.3	5.3	−	3.4	−	−	11.3	−	13.6	−	−	3.8	14.7	7.0	7.0	6.8	4.7
3	−	−	2.2	1.7	2.1	−	−	18.4	−	13.1	−	−	1.3	12.7	12.0	14.5	12.6	5.7

i14 = 12-methyl tridecanoic acid; 14:0 = tetradecanoic acid; i15 = 13-methyltetradecanoic acid; ai15 = 12-methyltetradecanoic acid; 15:0 = pentadecanoic acid; i16 = 14-methylpentadecanoic acid; 16:1 = hexadecanoic acid; 16:0 = hexadecanoic acid; i17 = 15-methylhexadecanoic acid; ai17 = 14-methylhexadecanoic acid; 17:0 = heptadecanoic acid; i18 = 16-methylheptadecanoic acid; 18:1 = octadecanoic acid; 18:0 = octadecanoic acid. a,b,c,d,e are unidentified peaks.
All of the strains contained both branched and straight chain fatty acids, but straight chained acids predominated.

Table 10. Amino acid composition of cell walls of Gram-positive filamentous carboxydotrophs.

Strain no.	Ala	Gly	Val	Thr	Ser	Leu	is-Leu	Pro	Dab	Asp	Phe	Orn	Glu	Lys	DapLL	Dapmeso
1	1.6	0.6	0.5	0.3	0.2	0.4	0.2	0.2	−	−	0.5	0.2	1.0	0.2	−	0.8
2	1.3	0.1	0.1	−	−	0.1	−	−	−	−	0.3	0.1	1.0	0.1	−	1.2
3	1.8	0.4	0.4	0.1	0.1	0.3	0.1	0.1	−	−	0.4	0.2	1.0	0.1	−	1.1

Abbreviations: Ala, alanine; Gly, glycine; Val, valine; Thr, threonine; Ser, serine; Leu, leucine; is-Leu, iso-leucine; Pro, proline; Dab, diaminobutyric acid; Asp, aspartic acid; Phe, phenylalanine; Orn, ornithine; Glu, glutamic acid; Lys, lysine; Dap, diaminopimelic acid; Dapmeso, meso isomer of diaminopimelic acid. The strains contained the meso isomer of Dap and had insignificant amounts of glycine present.

(i)   Production of bulk products such as single cell protein or PHB.

(ii)   Production of carbon monoxide sensors.

(iii)   Production of a carbon monoxide filter.

(iv)   Production of high value products from novel Gram-positive carboxydotrophic bacteria.

Single cell protein or PHB

Gases derived from coal, lignite or peat are relatively cheap and are therefore attractive for the production of single cell protein or PHB. They usually contain H_2 and CO and both gases may be used as energy sources for bacterial growth. Although the hydrogen bacteria have been considered as potential sources of single cell protein (Schlegel and Lafferty, 1971), the slow growth rates of these bacteria, the poor solubility of the gas and its explosive nature appear to preclude serious consideration.

Carboxydobacteria are more suitable for the conversion of synthesis gas to single cell protein since they can use either the hydrogen or CO components of the gas (Meyer, 1980, 1981). The growth rates of the mesophilic species are too low, but the thermophilic species have doubling times of approximately 3 h on CO. However, the toxic and explosive nature of synthesis gas and the poor solubility of hydrogen and carbon monoxide make it unlikely that synthesis gas will be used in this way. The production of PHB may be more attractive.

CO biosensors

CO oxidase from *P. thermocarboxydovorans* has no direct electroactivity with electrodes but can be coupled via mediators such as horse heart cytochrome *c*, phenazine ethosulphate or ferrocene derivatives (Turner *et al.*, 1984, 1985). Several CO biosensor designs have been investigated and the final design consisted of a platinum disc coated with the insoluble ferrocene derivative 1,1'-dimethylferrocene with CO oxidase retained at the surface of the electrode behind a gas-permeable membrane. A cylindrical piece of silver foil was used as a pseudo-reference electrode and a potential of $+150$ mV versus Ag/AgCl was applied to the electrode. Such probes responded rapidly and quantitatively to CO supplied as a gas or in solution.

CO biofilters

The use of a 'CO monoxidase' for the removal of toxic CO from a gas stream was patented in the USA by Scheinberg (1972). The enzyme was to be obtained from red blood cells, bacteria or plants. Major problems for most applications would appear to be:

(i)   the volume of catalyst required;
(ii) the harsh conditions to which the biocatalyst would be exposed; and
(iii) filter design with particular respect to contact area and the maintenance of the aqueous environment required for catalytic activity.

In our laboratories, we are attempting to solve some of these basic problems by applying modern techniques in molecular genetics and enzyme technology to a model filtration system based on the CO oxidase from *P. thermocarboxydovorans*.

Production of high value products from novel Gram-positive carboxydotrophic bacteria

Three new strains of Gram-positive, filamentous bacteria have been isolated on synthesis gas. All use hydrogen as energy source and fix CO_2 via the ribulosebisphosphate carboxylase cycle. The cell walls of the strains contain menaquinones with nine isoprene units (*Table 8*) and mainly straight chained fatty acids (*Table 9*). They also contain meso-diaminopimelic acid in their cell walls, together with alanine and glutamic acid and lack galactose (*Table 10*). These properties are characteristic of wall type II strains (C.Todd and C.Falconer, unpublished data). The strains have a G+C content of $75.2 - 75.4$ mol%. These actinomycetes and the previously described *Bacillus* strains may find useful applications for the production of enzymes, insecticides and pharmaceuticals.

References

Bamforth,C.W. and Quayle,J.R. (1977) Hydroxypyruvate reductase activity in *Paracoccus denitrificans*. *J. Gen. Microbiol.*, **101**, 259−267.

Baron,G., Bieger,F., Cornils,B., Franzen,U.E., Goeke,E.K., Lohmann,C., Tanz,H. and Traenckner, K.-C. (1982) Gasification of coal. In *Chemical Feedstocks from Coal*. Falbe,J., (ed.), John Wiley and Sons, New York, pp. 164−247.

Bell,J.M., Williams,E. and Colby,J. (1985) Carbon monoxide oxidoreductases from thermophilic bacteria. In *Microbial Gas Metabolism*. Poole,R.K. and Dow,C.S. (eds), SGM Special Publication no. 14, Academic Press, London, pp. 153−159.

Beudeker,R.F., Kuenen,J.G. and Codd,G.A. (1981) Glycollate metabolism in the obligate chemolithotroph *Thiobacillus neapolitanus* grown in continuous culture. *J. Gen. Microbiol.*, **126**, 337−346.

Colby,J., Williams,E. and Turner,A.P.F. (1985) Applications of CO-utilising microorganisms. *Trends Biotechnol.*, **3**, 12−27.

Cypionka,H. and Meyer,O. (1982) Influence of carbon monoxide on growth and respiration of carboxydobacteria and other aerobic organisms. *FEMS Microbiol. Lett.*, **15**, 209−214.

Cypionka,H. and Meyer,O. (1983) CO-insensitive respiratory chain of *Pseudomonas carboxydovorans*. *J. Bacteriol.*, **156**, 1178−1187.

Frohning,C.D., Kolbel,H., Ralek,M., Rottig,W., Schnur,F. and Schulz,H. (1982) Fischer-Tropsch process. In *Chemical Feedstocks from Coal*. Falbe,J. (ed.), John Wiley and Sons, New York, pp.309−432.

Gibbs,D.F. and Greenhalgh,M.E. (1983) *Biotechnology, Chemical Feedstocks and Energy Utilisation*. Frances Pinter, London.

Graboski,M.S. (1984) The production of synthesis gas from methane, coal and biomass. In *Catalytic Conversions of Synthesis Gas and Alcohols to Chemicals*. Herman,R.G. (ed.), Plenum Press, New York, pp.37−52.

Harwood,J.H., Williams,E. and Bainbridge,B.W. (1972) Mutation in the methane-oxidising bacterium, *Methylococcus capsulatus*. *J. Appl. Bacteriol.*, **35**, 99−108.

Hughes,M.N. (1985) The inorganic chemistry of microbial gas metabolism. In *Microbial Gas Metabolism. Mechanistic, Metabolic and Biotechnological Aspects*. Poole,R.K. and Dow,C.S. (eds), Academic Press, London, pp. 3−30.

Kesler,T.G., Stasishina,G.N., Kasaeva,G.E. and Sid'ko,F.Y.A. (1978) Possibility of culturing hydrogen bacteria in the presence of CO. *Microbiology*, **47**, 17−20.

Kim,Y.M. and Hegeman,G.D. (1983) Oxidation of carbon monoxide by bacteria. *Int. Rev. Cytol.*, **81**, 1−32.

Knifton,J.F., Grigsby,R.A. Jr, and Herbstman,S. (1984) Alcohol/ester fuels from synthesis gas. In *Catalytic Conversions of Synthesis Gas and Alcohols to Chemicals*. Herman,R.G. (ed.), Plenum Press, New York, pp. 81−96.

Kölling,G. and Schnur,F. (1982) In *Chemical Feedstocks from Coal*. Falbe,J. (ed.), John Wiley and Sons, New York, pp 1−11.

Kruger,B. and Meyer,O. (1984) Thermophilic *Bacilli* growing with carbon monoxide. *Arch. Microbiol.*, **139**, 402−408.

Lyons,C.M., Justin,P., Colby,J. and Williams,E. (1984) Isolation, characterisation and autotrophic metabolism of a moderately thermophilic carboxydobacterium, *Pseudomonas thermocarboxydovorans* sp. nov. *J. Gen. Microbiol.*, **130**, 1097-1105.

Meyer,O. (1980) Using carbon monoxide to produce single-cell protein. *Bioscience*, **30**, 405−407.

Meyer,O. (1981) Growth of carbon monoxide oxidising bacteria with industrial gas mixtures, automobile exhaust gas and other unconventional CO-containing gases. *Stud. Environ. Sci.*, **9**, 79−86

Meyer,O. and Schlegel,H.G. (1980) Carbon monoxide: methylene blue oxidoreductase from *Pseudomonas carboxydovorans*. *J. Bacteriol.*, **141**, 74−80.

Meyer,O. and Schlegel,H.G. (1983) Biology of aerobic carbon monoxide-oxidising bacteria. *Annu. Rev. Microbiol.*, **37**, 277−310.

Meyer,O. and Rohde,M. (1984) Enzymology and bioenergetics of carbon monoxide-oxidising bacteria. In *Microbial Growth on C1 Compounds*. Crawford,R.L. and Hansom,R.S. (eds), American Society for Microbiology, Washington, DC, pp. 26−33

Nelder,J.A. and Mead,R. (1965) A simplex method for function minimisation. *Computer J.*, **7**, 308.

Scheinberg,I.H. (1972) Filters and carbon monoxide indicators. US Patent no. 3,6983,327.

Schlegel,H.G. and Lafferty,R.M. (1971) The production of biomass from hydrogen and carbon dioxide. *Adv. Biochem. Eng.*, **1**, 143−168.

Seglin,L., Geosits,R., Franko,B.R. and Gruber,G. (1975) Survey of methanation chemistry and processes. In *Methanation of Synthesis Gas. Advances in Chemistry Series 146*. American Chemical Society, Washington, DC.

Sheldon,R.A. (1983) *Chemicals from Synthesis Gas*. D. Reidel Publishing Company, The Netherlands.

Taylor,S.C. (1977) Evidence for the presence of ribulose 1,5-bisphosphate carboxylase and phosphoribulokinase in *Methylococcus capsulatus* (Bath). *FEMS Microbiol. Lett.*, **2**, 305−307.

Turner,A.P.F., Aston,W.J., Higgins,I.J., Bell,J.M., Colby,J., Davis,G. and Hill,H.A.O. (1984) Carbon monoxide:acceptor oxidoreductase from *Pseudomonas thermocarboxydovorans* strain C2 and its use in a carbon monoxide sensor. *Anal. Chim. Acta*, **163**, 161−174.

Turner,A.P.F., Aston,W.J., Davis G., Higgins,I.J., Hill,H.A.O. and Colby,J. (1985) Enzyme-based carbon monoxide sensors. In *Microbial Gas Metabolism. Mechanistic, Metabolic and Biotechnological Aspects.* Poole,R.K. and Dow,C.S. (eds), SGM Special Publication no. 14, Academic Press, London, pp. 161−170.

van Heek,K.H. and Juntgen,H. (1982) Production of synthesis gas and methane via coal gasification utilising nuclear heat. In *Chemical Feedstocks from Coal.* Falbe,J. (ed.), John Wiley and Sons, New York, pp.272−308.

Volova,T.G., Kalcheva, G.S., Stasishina, G.N. and Kasaeva, G.E. (1980) Growth of hydrogen bacteria during inhibition by carbon monoxide. *Microbiology*, **49**, 465−471.

Williams,E. and Bainbridge,B.W. (1976) Mutation, repair mechanisms and transformation in the methane-oxidising bacterium, *Methylococcus capsulatus. Proceedings of the 2nd International Symposium on the Genetics of Industrial Microorganisms.* MacDonald,K.D., (ed.), Academic Press, New York, pp. 313−322.

Williams,E. and Colby,J. (1986) Biotechnological applications of carboxydotrophic bacteria. *Microbiol. Sci.*, **3**, 149−153.

Williams,E. and Shimmin,M.A. (1978) Radiation induced filamentation in obligate methylotrophs. *FEMS Microbiol. Lett.*, **4**, 137−141,

Williams,E., Shimmin,M.A. and Bainbridge,B.W. (1977) Mutation in the obligate methylotrophs, *Methylococcus capsulatus* and *Methylomonas albus. FEMS Microbiol. Lett.*, **2**, 293−296.

Williams,E., Colby,J., Lyons,C.M., and Bell,J. (1986) The bacterial utilisation of synthetic gases containing carbon monoxide. *Biotechnol. Genet. Eng. Rev.*, **4**, 169−211.

Concluding Remarks

J.R.QUAYLE

University of Bath

Conferences may be classified into three broad types: those which bring together experts in a narrow subject area to discuss in detail the results of recent research; those which address wide audiences at a general level; those which mix experts in different but related fields in order to inform them of developments in separate areas which are interacting or likely to interact in the near future. The present conference has been of the last type and has brought together very successfully research leaders in biochemistry and microbial physiology, biotechnologists and commodity experts from the European Commission and related bodies.

It is as well to remember that biotechnology, like any other manufacturing industry, is all about converting a chemical feedstock, be it a pure compound or a mixture of compounds, into a product or products which command a sufficiently higher price in the market to pay for the cost of the value-adding process, leaving a profit margin over for reinvestment in the company and distribution to the shareholders. Additionally, employment is generated and the country's wealth is increased through taxes. The profit margin is the *sine qua non* and lack of it can only be carried on the back of a more successful operation.

Until about thirty years ago, biotechnology was a somewhat inexact science involving certain well established fermentations, processes of antibiotic production, and micro-biological operations mainly in the food, drink and dairy industries. It has now become a very sophisticated science critically dependent on advanced basic research in the biological sciences and in process engineering. As the chapters by C.Anthony (Chapter 6) and S.H.Collins (Chapter 9) showed, it is now possible to calculate fairly precisely the maximum yield data for transformation of a chemical feedstock into a product. In practice, the theoretical maximum will not be attained due to energy 'wastage' by the organism operating under, for example, conditions of substrate excess or high growth rate. Such factors were brought out by M.J.Teixeira de Mattos (Chapter 10) in his chemostat study of the growth of *Klebsiella aerogenes* on a range of carbohydrate substrates. Such detailed knowledge of the growth physiology of an organism is essential for the design of an economic fermentation process. Part of the energy budget of many fermentations is the cost of transport of the substrate into the cell and P.J.F.Henderson (Chapter 5) demonstrated the enormous power of molecular genetics in dissecting such processes to the molecular level, using carbohydrate transport into *Escherichia coli* as an example.

It should not be forgotten that entirely new possibilities in biotechnology can be opened up with discovery of a new class of organisms; this was clearly illustrated by E.Williams

(Chapter 11) who described the rapid progress made with the carboxydotrophs, including the whole range of new CO-utilizing actinomycetes.

There are now huge gluts in the world production of hydrocarbons and agricultural products such as carbohydrates, carbohydrate polymers, oils and fats and several contributors between them [Coombs, Drozd, Woodward and Stowell (see Chapters 3, 7, 4 and 8)] surveyed the use of these as fermentation feedstocks. Two things were abundantly clear: firstly, wherever possible, flexibility in the starting feedstock is of great value in adjusting to changing availability and price on the world market; secondly, despite the availability of large scale plant, biotechnology still awaits a major new process for profitable manufacture of a high volume, low cost product; it is still mainly concerned with low volume, high cost products.

The appearance of a new high volume process is eagerly awaited by the oil companies and by the growers in the EEC. Huge surpluses of agricultural products continue to grow in the EEC and there is an urgent need to find new uses for them in the chemical and biotechnological industries. The two contributors who represented the European growers (Cormack) and the Commission (Gray) detail the problems in Chapters 2 and 1 respectively, including the complexities of pricing regulations. In view of the continually rising levels of agricultural productivity and the expected impact of genetically engineered crops in about five years' time, it is clear that these surpluses are going to remain a permanent feature for the foreseeable future and a target area for new processes.

INDEX